Proton-Coupled Electron Transfer
A Carrefour of Chemical Reactivity Traditions

RSC Catalysis Series

Series Editor:
Professor James J Spivey, *Louisiana State University, Baton Rouge, USA*

Advisory Board:
Krijn P de Jong, *University of Utrecht, The Netherlands,* James A Dumesic, *University of Wisconsin-Madison,* USA, Chris Hardacre, *Queen's University Belfast, Northern Ireland,* Enrique Iglesia, *University of California at Berkeley,* USA, Zinfer Ismagilov, *Boreskov Institute of Catalysis, Novosibirsk, Russia,* Johannes Lercher, *TU München, Germany,* Umit Ozkan, *Ohio State University, USA,* Chunshan Song, *Penn State University, USA*

Titles in the Series:
1: Carbons and Carbon Supported Catalysts in Hydroprocessing
2: Chiral Sulfur Ligands: Asymmetric Catalysis
3: Recent Developments in Asymmetric Organocatalysis
4: Catalysis in the Refining of Fischer–Tropsch Syncrude
5: Organocatalytic Enantioselective Conjugate Addition Reactions: A Powerful Tool for the Stereocontrolled Synthesis of Complex Molecules
6: N-Heterocyclic Carbenes: From Laboratories Curiosities to Efficient Synthetic Tools
7: P-Stereogenic Ligands in Enantioselective Catalysis
8: Proton-Coupled Electron Transfer: A Carrefour of Chemical Reactivity Traditions

How to obtain future titles on publication:
A standing order plan is available for this series. A standing order will bring delivery of each new volume immediately on publication.

For further information please contact:
Book Sales Department, Royal Society of Chemistry, Thomas Graham House, Science Park, Milton Road, Cambridge, CB4 0WF, UK
Telephone: +44 (0)1223 420066, Fax: +44 (0)1223 420247, Email: books@rsc.org
Visit our website at http://www.rsc.org/Shop/Books/

Proton-Coupled Electron Transfer
A Carrefour of Chemical Reactivity Traditions

Edited by

Sebastião Formosinho
Department of Chemistry, University of Coimbra, Portugal

Mónica Barroso
Department of Chemistry, Imperial College London, UK

RSC Publishing

RSC Catalysis Series No. 8

ISBN: 978-1-84973-141-6
ISSN: 1757-6725

A catalogue record for this book is available from the British Library

© Royal Society of Chemistry 2012

Published by The Royal Society of Chemistry,
Thomas Graham House, Science Park, Milton Road,
Cambridge CB4 0WF, UK

Registered Charity Number 207890

For further information see our web site at www.rsc.org

Preface

The topic of "proton-coupled electron transfer" (PCET) has received increasing attention in the last decades, partly due to the realization of its role in the context of important biological and chemical (catalytic) processes. The mechanisms of vital functions like respiration and photosynthesis, and in general most of the biological enzyme processes, are now known to benefit from combining proton and electron movements in many of the charge transfer steps. The level of understanding of such systems and processes has increased rapidly in recent times, as a consequence of advances in characterisation techniques, particularly X-ray crystallography, time-resolved spectroscopies and electrochemistry, together with advances in theory and computational methods. Interestingly, the concept of PCET itself has been the subject of some dispute, with several groups disagreeing on the significance of that designation. A reasonably consensual definition of PCET is the combined transfer of protons and electrons from different sites and/or to different sites of the system, resulting in the overall transfer of an hydrogen atom, in opposition to the conventional hydrogen atom transfer (HAT). Further mechanistic considerations can then be added, to distinguish between concerted and stepwise processes and, in the latter case, the ones which start with the transfer of the proton and those where proton transfer follows the electron transfer. A better knowledge of the thermodynamic and kinetic aspects of these reactions can help clarify the mechanistic differences and contribute to further develop relevant theories and models, ultimately guiding the design of artificial systems, particularly relevant in the contexts of enzyme catalysis and energy conversion.

Chemical reactivity is currently explained in terms of several scientific traditions. One of them is the bond-breaking–bond-forming process and conceptually based on Potential Energy Surfaces. A second one is based on the role of Franck–Condon factors due to the overlap of vibrational wavefunctions.

RSC Catalysis Series No. 8
Proton-Coupled Electron Transfer: A Carrefour of Chemical Reactivity Traditions
Edited by Sebastião Formosinho and Mónica Barroso
© Royal Society of Chemistry 2012
Published by the Royal Society of Chemistry, www.rsc.org

Quantum mechanical tunnelling of vibrational modes is an equivalent way to account for such factors. A third tradition has emerged from a proposal of Prof. Rudolph Marcus on the role of the solvent configuration around a charged species, the so-called solvent reorganization. PCET is a "Carrefour" of such traditions, together with quantum mechanical tunnelling of the electron particle. Therefore, a purpose of the book is, through personal accounts, to bring together such traditions, which in science have always an intention of universality due to the contact of the scientist with reality through the experimental observations. Are all the abovementioned traditions compatible with each other and with the present level of experimental observation in the field? Chapters 1 and 2 of this book aim to give some insight on this problem, discussing the problem of PCET and HAT in the frameworks of Marcus theory and the Intersecting/Interacting State Model, respectively.

From the experimental point of view, the role of PCET in biological or biologically inspired systems will be explored. Chapter 3 presents an overview of theoretical and experimental techniques used to analyse PCET and their application in practical examples of enzyme reactions. The last chapters focus on two other important biological functions where PCET plays a central role, respiration (Chapter 4) and photosynthesis (Chapter 5), and how the design principles used by nature to optimise and regulate these processes can be a useful guide to the design of artificial systems, particularly in the context of fuel cells and artificial photosynthesis/solar fuel production. We hope that these accounts will bring together what have been some of the most recent developments in the topical subject of PCET, and provide the reader with an insight in the current understanding and applications of this important type of reactions.

Sebastião Formosinho
Mónica Barroso

Contents

RSC Catalysis Series No. 8
Proton-Coupled Electron Transfer: A Carrefour of Chemical Reactivity Traditions
Edited by Sebastião Formosinho and Mónica Barroso
© Royal Society of Chemistry 2012
Published by the Royal Society of Chemistry, www.rsc.org

CHAPTER 1

Application of the Marcus Cross Relation to Hydrogen Atom Transfer/Proton-Coupled Electron Transfer Reactions

JEFFREY J. WARREN[†] AND JAMES M. MAYER*

Department of Chemistry, University of Washington, Box 351700, Seattle, WA 98195-1700, USA

1.1 Introduction

Many important chemical and biological reactions involve transfer of both electrons and protons.[1] This is illustrated, for instance, by Pourbaix's extensive 1963 Atlas of Electrochemical Equilibria.[2] These have come to be called 'proton-coupled electron transfer' (PCET) reactions.[3–5] Due to the widespread interest in this topic, the term PCET is being used by many authors in a variety of different contexts and with different connotations. As a result, a very broad definition of PCET has taken hold, encompassing any redox process whose rate or energetics are affected by one or more protons. This includes processes in which protons and electrons transfer among one or more reactants, regardless of mechanism, and processes in which protons modulate ET processes even if they do not transfer.[6]

[†]Current Address: Department of Chemistry, California Institute of Technology, 1200 E. California Boulevard, Pasadena, CA 91125, USA

RSC Catalysis Series No. 8
Proton-Coupled Electron Transfer: A Carrefour of Chemical Reactivity Traditions
Edited by Sebastião Formosinho and Mónica Barroso
© Royal Society of Chemistry 2012
Published by the Royal Society of Chemistry, www.rsc.org

Mechanistic issues are central to PCET. In contrast to electron transfer (ET) and proton transfer (PT), which are two of the most fundamental and well-understood reactions in chemistry, our understanding of how protons and electrons are transferred *together* is still emerging. The importance of mechanism was emphasized by Njus in a biochemical context almost two decades ago: "Many [biological redox] reactions involve the transfer of hydrogen atoms (or the concerted transfer of H^+ and e^-) rather than electron transfer alone. This distinction is generally disregarded because $H^\bullet$ and e^- are considered interchangeable in the aqueous milieu of the cell, but the focus on electrons obscures some of the general principles underlying the functioning of redox chains".[7]

This chapter focuses on hydrogen atom transfer (HAT) reactions, which involve concerted transfer of a proton and an electron from a single donor to a single acceptor in one kinetic step (eqn (1.1)). These are one subset of PCET processes and are one type of 'concerted proton-electron transfer' (CPET).[8]

$$X\text{–}H + Y \xrightarrow{k_{XH/Y}} X + Y\text{–}H \tag{1.1}$$

"Concerted" implies a single kinetic step for transfer of the two particles, but does not imply synchronous transfer. HAT is a fundamental reaction studied by physical and organic chemists for over a century, critical to combustion and free-radical halogenations, for example.[9] More recently, it has been recognized that transition metal coordination complexes and metalloenzymes can undergo HAT reactions, and the recognition of overlap between traditional HAT reactions and PCET has stimulated much new thinking.[10–13] Our focus has been to understand the key factors that dictate HAT and PCET reactivity and to build a simple and predictive model that can be used in chemistry and in biology.[5]

In this chapter, we show that the Marcus cross relation holds remarkably well for HAT reactions in most cases. This provides important insights into HAT and allows the prediction of rate constants. We begin with an introduction to Marcus theory and the cross relation. This is followed by applications the cross relation to purely organic reactions (Sections 1.3),[14] and then to HAT reactions involving transition metal complexes (Sections 1.4).[15] Finally, Section 1.5 describes the intuitive picture of HAT derived from the success of the cross relation, and also emphasizes some of the weaknesses of this treatment and the questions that remain.

1.2   An Introduction to Marcus Theory

The Marcus theory of electron transfer has proven invaluable for understanding a variety charge transfer reactions, from simple solution reactions to long-range biological charge transfer.[16–19] The primary equation of Marcus theory, equation (1.2), is derived from a model of intersecting parabolic free energy surfaces.[19] When the coupling between these diabatic surfaces H_{AB} is small, the reaction is non-adiabatic and the reaction does not always occur

$$k_{et} = \frac{2\pi}{\hbar} H^2{}_{AB} \frac{1}{\sqrt{4\pi\lambda RT}} \exp\left(-\frac{(\Delta G^\circ + \lambda)^2}{4\lambda RT}\right) \quad (1.2)$$

$$k_{et} = A e^{-(\Delta G^\circ + \lambda)^2/4\lambda RT} \quad (1.3)$$

when the system reaches the intersection (the transition state). When the coupling is sufficiently large the reaction is adiabatic and equation (1.2) reduces to equation (1.3). The pre-exponential factor A in equation (1.3), for a bimolecular reaction, is typically taken as an adjusted collision frequency. The intrinsic barrier λ is the energy required to distort the reactants and their surrounding solvent to the geometry of the products. Because electron transfer occurs over relatively long distances, with little interaction between the reagents, it is typically assumed that λ can be taken as a property of the individual reagents. λ for a reaction is then commonly taken as the average of the individual reagent λ's (the 'additivity postulate,' eqn (1.4)). In the adiabatic limit, λ for an individual reagent can be determined from the rate of the self-exchange reaction (eqn (1.5)). Combining equations (1.3) and (1.4) gives the cross relation (eqn (1.6) and (1.7)), which relates the rate constant of a cross reaction, $X + Y^-$, to the self exchange rate constants for reagents X and Y (eqn (1.5)) and the equilibrium constant K_{XY}. The constant f is defined by equation (1.7) and is typically close to unity, unless $|\Delta G^\circ| \geq \lambda/4$.[17]

$$\lambda_{XY} = \frac{1}{2}(\lambda_{XX} + \lambda_{YY}) \quad (1.4)$$

$$X + X^- \xrightarrow{k_{XX}} X^- + X \quad (1.5)$$

$$k_{XY} = \sqrt{k_{XX} k_{YY} K_{XY} f} \quad (1.6)$$

$$\ln f = \frac{(\ln K_{XY})^2}{4\ln(k_{XX} k_{YY} Z^{-2})} \quad (1.7)$$

Theoretical treatments of PCET reactions typically have equation (1.2) as a conceptual starting point. In Hammes–Schiffer's multistate continuum theory for PCET,[13] the pre-exponential factor includes both electronic coupling and vibrational overlaps, and the rate is a sum over initial and final vibrational states integrated over a range of proton-donor acceptor distances. This theory has been elegantly applied to understand the intimate details of a variety of PCET reactions, but many of its parameters are essentially unattainable experimentally.

The cross relation can be written for an HAT reaction (eqn (1.1) and (1.8)). It is a very simplistic model, but it has the advantage that all of the parameters are experimentally accessible (in many cases).

$$k_{XH/Y} = \sqrt{k_{XH/X}\, k_{YH/Y}\, K_{XH/Y} f} \qquad (1.8)$$

It should be emphasized that the cross relation is not a corollary of current PCET theory and that there is little theoretical justification for applying it (although Marcus has briefly discussed this).[20] Still, the cross relation has been successfully applied to group transfer reactions including proton[21] and hydride transfers,[22] and S_N2 reactions.[23] While these successes are notable, in each instance the cross relation holds only over a narrow set of reactants and reactions. In contrast, the treatment described here has shown to be a powerful predictor for a wide array of HAT reactions.

Our interest in applying the Marcus cross relation grew out of our finding that the traditional Bell–Evans–Polanyi (BEP) relationship, $E_a = \alpha(\Delta H) + \beta$,[9,24] holds well for transition metal complexes abstracting hydrogen atoms from C–H bonds.[25] The BEP equation relates HAT activation energies to the enthalpic driving force (ΔH) (although, as discussed in Section 1.4 below, free energies should be used, as in Marcus theory). The ΔH is typically taken as the difference in bond dissociation enthalpies (BDEs) of X–H and Y–H.[26] The BEP equation has been a cornerstone of organic radical chemistry for many decades, typically holding well for reactions of one type of oxidant $X^\bullet$ with a series of substrates Y–H. The success of this treatment is one reason why organic textbooks list BDEs.[27] We initially found that the rate constants for HAT from C–H bonds to CrO_2Cl_2 or MnO_4^- show good BEP correlations with the BDE of the C–H bond.[28] Later, we found an excellent BEP correlation for C–H bond oxidations by $[Ru(O)(bpy)_2(py)]^{2+}$ (Figure 1.1).[29] Such a correlation, with a Brønsted slope $\Delta\Delta G^{\ddagger}/\Delta\Delta H^\circ$ close to $\frac{1}{2}$, is a strong indicator of an HAT mechanism. Many other groups have also used these correlations to understand the relationship between rate and driving force for HAT reactions of transition metal containing systems.[30] Marcus theory and the cross relation also predict a Brønsted slope ($\Delta\Delta G^{\ddagger}/\Delta\Delta G^\circ$) close to $\frac{1}{2}$, for reactions at low driving force (specifically when $\Delta G^\circ \ll \lambda/2$).

The BEP correlation between rates and driving force for HAT is very valuable, but it applies only to a specific set of similar reactions, for instance MnO_4^- abstracting $H^\bullet$ from hydrocarbons.[31] In addition, the α and β parameters are defined only with the context of the correlation and have no independent meaning. In contrast, cross relation uses three independently measurable parameters: the equilibrium constant $K_{XH/Y}$ (which is equal to $e^{-\Delta G^\circ_{XH/Y}/RT}$) and the rate constants for the hydrogen atom self-exchange reactions $k_{XH/X}$ and $k_{YH/Y}$ (eqn 1.9).

$$XH + X \rightarrow X + XH \qquad (1.9)$$

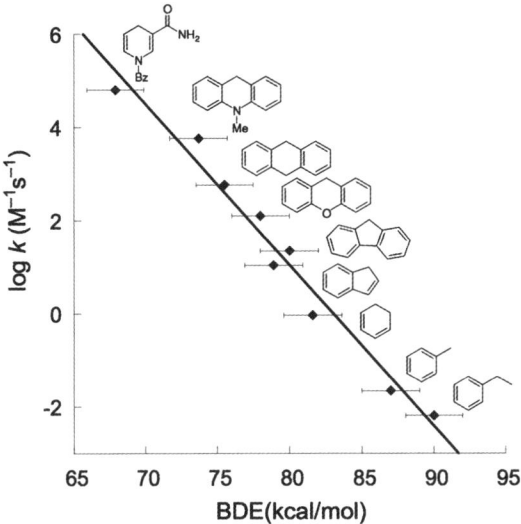

Figure 1.1 Plot of statistically corrected rate constants *versus* BDE for H-abstraction from C–H bonds by [Ru(O)(bpy)$_2$py]$^{2+}$.[29]

1.3 Predicting Organic Hydrogen Atom Transfer Rate Constants

Hydrogen atom transfer (HAT) reactions of organic compounds are fundamental to combustion, industrial oxidation processes, and biological free radical chemistry, among other areas of chemistry and biology. One important example is the series of H-transfers that is thought to be involved in lipid oxidation. Peroxyl radicals (ROO$^{\bullet}$) abstract H$^{\bullet}$ from a lipid to give a lipid radical that adds O$_2$ to form a new peroxyl radical and propagate the radical chain.[32] ROO$^{\bullet}$ can also abstract H$^{\bullet}$ from α-tocopherol (a component of vitamin E) and the resulting α-tocopheroxyl radical is thought to be regenerated *via* HAT from ascorbate (vitamin C).[32] Understanding such a web of free radical reactions requires knowledge of the rate constants for each of the steps. To this end, we have developed a predictive model for organic HAT reactions[14] based upon the Marcus cross relation and the kinetic solvent effect model of Ingold *et al.*[33]

We begin this section discussing the application of the cross relation to real systems, how the needed rate and equilibrium constants can be obtained. These same principles also apply to the metal-mediated HAT reactions discussed in Section 1.4. A set of reactions are used to test the Marcus model, using inputs all obtained in the same solvent. Then we address how to extrapolate rate and equilibrium constants from one solvent to another, using the H-bonding descriptors developed by Abraham and co-workers.[34–36] Finally, we show that this allows remarkably accurate prediction of a very wide range of HAT cross rate constants.[14]

1.3.1 Obtaining Self-Exchange Rate Constants and Equilibrium Constants

Ideally, all three of the parameters needed for the cross relation, $K_{XH/Y}$, $k_{XH/X}$ and $k_{YH/Y}$, are measured in the same medium under the same conditions. When the values are only available in different solvents, solvent corrections must be included, as described in Section 1.3.3 below. The f term can be calculated from the three parameters, with the collision frequency Z typically taken as 10^{11} M^{-1} s^{-1}.[37,38]

The driving force for a HAT reaction, $\Delta G^{\circ}_{XH/Y} = -RT \ln K_{XH/Y}$, is best determined by direct equilibrium measurements in the solvent of interest. However, this is typically limited to reactions where $|\Delta G^{\circ}_{XH/Y}|$ is small, less than about 5 kcal mol^{-1}. Also, this is only possible for reactions in which all of the species are fairly stable, which is unusual for organic radical reactions. The ΔG° for a HAT reaction is typically more easily derived as the difference in bond dissociation free energies (BDFEs) of X–H and Y–H in the solvent of interest. We have recently reviewed BDFEs of common organic and biochemical species and how they are obtained,[39] so only an overview is given here.

One powerful method to determine BDFEs uses a solution thermochemical cycle with the reduction potential of XH and the pK_a of XH$^+$, or with $E^{\circ}(X^-)$ and $pK_a(XH)$. The BDFE in kcal mol^{-1} is given by $23.1E^{\circ} + 1.37pK_a + C_G$.[39–42] Bordwell and others have used this approach to measure many bond dissociation *enthalpies* (BDEs)[40] but it is more appropriate to use BD*FE*s because the E° and pK_a values are free energies.[39,41,42] Determining X–H BD*E*s from E° and pK_a measurements is valid when XH and X have similar absolute entropies, as is typically the case for organic molecules but not for transition metal complexes (see Section 1.4.1 below).[39,41,42] Due to the uncertainties in the C_G value in thermochemical cycle, and typical uncertainties in the E° and pK_a values, this procedure yields BDFEs accurate to no better than ± 1 kcal mol^{-1}. This leads to estimated uncertainties in rate constants calculated from the cross relation of an order of magnitude.

Solution-phase BDFEs can also be obtained from gas-phase BDEs, which are available for many small organic molecules. An extensive tabulation of such BDEs can be found in the recent book by Luo, portions of which are available online.[43] As described in detail elsewhere,[14,39] a gas-phase BDE can be converted into the corresponding solution-phase BDFE using data from standard tables [$S^{\circ}(H^{\bullet})_{gas}$, $\Delta G^{\circ}_{solvation}(H^{\bullet})$] and an estimate of the difference in the free energies of solvation of XH and X (see below).

Self-exchange rate constants, $k_{XH/X}$ and $k_{YH/Y}$, are best measured directly when this is possible. NMR line broadening is a powerful technique for degenerate exchange reactions of stable species if the rate constant is *ca.* 10^3–10^6 M^{-1} s^{-1};[44] faster reactions can be monitored by EPR methods.[45] In the ^{1}H NMR experiments, typically one reactant is diamagnetic and has a sharp spectrum while the other is paramagnetic. In the slow-exchange limit, addition of the paramagnetic species to the diamagnetic causes broadening of the spectrum but not shifting, and the amount of broadening is directly related to the rate constant. We have used this method to measure a number of $k_{XH/X}$ values for transition metal reagents.[29b,46–49]

Self-exchange rate constants can also be determined through the use of 'pseudo-self-exchange' reactions, that is H• exchange reactions using two very similar reagents X(H) and *X(H) (eqn (1.10)). The reagents can differ in just an isotopic label (*e.g.* toluene/3-deuterotoluene)[50] or just be chemically similar. For instance, we have examined the pseudo-self-exchange reaction of oxo-vanadium complexes that differ only in their 4,4'-dimethylbipyridine *vs.* 4,4'-di-(*t*-butyl)bipyridine supporting ligands.[51] This reaction has $K_{XH/*X} = 1$ within experimental error, so it is very close to a true self-exchange reaction. Reaction of the hydroxylamine TEMPO-H (2,2,6,6-tetramethyl-*N*-hydroxypiperidine) with the aminoxyl radical 4-oxo-TEMPO (eqn (1.11)) has $K_{XH/*X} = 4.5 \pm 1.8$.[52] In such cases the self-exchange rate constant $k_{XH/X}$ is taken to be the geometric mean of the forward ($k_{XH/*X}$) and reverse ($k_{*XH/X}$) rate constants (eqn (1.12)).[53]

$$XH + {}^*X^{\bullet} \rightarrow X^{\bullet} + {}^*XH \tag{1.10}$$

$$k_{XH/X} = \sqrt{k_{XH/^*X} k_{^*XH/X}} = k_{XH/^*X}\left(K_{XH/^*X}\right)^{-1/2} \tag{1.12}$$

Using these various approaches, homolytic bond strengths and self-exchange rate constants have been derived for a number of reagents. A selection of those used in this chapter are summarized in Table 1.1;[54,55] a more complete list of BDEs and BDFEs is given in references 14 and 39.

1.3.2 Tests of the Cross Relation for Organic HAT Reactions

To test the applicability of the cross relation to HAT, a set of 17 organic reactions have been compiled in which cross and self-exchange rate constants have all been measured under similar conditions (the self-exchange rate for 9,10-dihydroanthracene (DHA) has been estimated by applying the cross relation).[51] These reactions, indicated with a * in Table 1.2,[56–68] involve oxyl radicals abstracting H• from O–H and C–H bonds. The equilibrium constants are either available under the same conditions or have been adjusted using the solvent corrections described below.

Table 1.1   Properties of reagents in selected solvents: solution bond dissociation free energies (BDFEs) and self-exchange rate constants ($k_{XH/X}$) in selected solvents.[a]

Compound (α_2^H)[b]	Solvent (β_2^H)[b]	BDFE[a]	$k_{XH/X}$[a]
$^tBu_3PhOH$ (0.2)[c]	MeCN (0.44)	77.8	2.0×10^1
BHT (0.2)[c,d]	MeCN (0.44)	77.5	2.0×10^1
2,6-$^tBu_2PhOH$ (0.2)[c]	Styrene (0.18)	78.5	6.0×10^1
$^tBu_2(MeO)PhOH$ (0.2)[c,e]	MeCN (0.44)	74.9	2.0×10^1
Phenol (0.60)	MeCN (0.44)	87.8	3.2×10^5
Hydroquinone (0.53)	MeCN (0.44)	78	1.6×10^5 [f]
Tyrosine (0.60)	Water (0.38)	87.8	6.4×10^4
1-Naphthol (0.61)	Isopentane (0)	79.7	9×10^5
2-Naphthol (0.61)	Isopentane(0)	83.0	$\leq 9 \times 10^5$
TEMPOH (0.39)	MeCN (0.44)	66.5	4.7
Et_2NOH (0.29)	MeCN (0.44)	72.0	2.0×10^1
$^tBu_2NOH$ (0.29)	MeCN (0.44)	64.5	2.0×10^1
^tBuOOH (0.44)	Hexane (0)	80.4	5.0×10^2
L-Ascorbate (0.3^g)	Water (0.38)	73.6	8×10^5
$iAscH^-$ (0.3 ^{g})[h]	MeCN (0.44)	66.4	5×10^5
α-Tocopherol (0.4)	Styrene (0.18)	74.0	1.5×10^5
Trolox C (0.4)[i]	Water (0.38)	78.5	3×10^5
^tBuOH (0.32)	DTBP (0.35)[g]	104.4	3×10^4
DHA (0)[j]	MeCN (0.44)	75	5×10^{-11} [k,l]
Xanthene (0)	MeCN (0.44)	73	1×10^{-10} [k,l]
Fluorene (0)	MeCN (0.44)	77	1×10^{-10} [k,l]
Toluene (0)	DTBP (0.35)[g]	86.8	8×10^{-5} [k]
$[Fe(H_2bip)_3]^{2+}$	MeCN (0.44)	66.2	1.8×10^3
$[Fe(H_2bim)_3]^{2+}$	MeCN (0.44)	71.7	9.7×10^2
$[(^tBu_2bpy)_2V(O)(OH)]^+$	MeCN (0.44)	70.6	6.5×10^3
$(acac)_2Ru(pyimH)$	MeCN (0.44)	62.1	3.2×10^5
$TpOs(NH_2Ph)Cl_2$	MeCN (0.44)	61.5	1.5×10^{-3}
$[(bpy)_2(py)RuOH]^{2+}$	Water (0.38)	84.8	–
$[(bpy)_2(py)RuOH]^{2+}$	MeCN (0.44)	83	7.6×10^4
$[(phen)_4Mn_2(O)(OH)]^{3+n}$	MeCN (0.44)	74.7	4×10^3 [l]
$[(phen)_4Mn_2(OH)_2]^{3+}$	MeCN (0.44)	70.0	3×10^5 [l]

[a] BDFE in kcal mol^{-1} and $k_{XH/X}$ in M^{-1} s^{-1} at 298 K. A more complete list, with full derivations for BDFEs, self-exchange rate constants and accompanying references, is given in reference 14. BDFEs are reported to one decimal place in most cases to eliminate ambiguity due to rounding. Values with only two significant figures have errors greater than ± 1 kcal mol^{-1}.[14] Self-exchange rate constants are given to one decimal place unless KSE[14] or cross relation[51] approximations introduce relatively large errors. Abbreviations for ligands of transition metal complexes are given in ref. 54.
[b] α_2^H(solute) from reference 35 and β_2^H(solvent) from reference 36. For saturated alkyl compounds $\alpha_2^H = \beta_2^H = 0$. α_2^H and β_2^H are not known for transition metal complexes.
[c] α_2^H and $k_{XH/X}$ for the 2,6-di-t-butyl-4-R substituted phenols are approximated as equal due to their structural similarity.
[d] BHT = 2,6-di-$tert$-butyl-4-methyl-phenol.
[e] $^tBu_2OMePhOH$ = 2,6-di-$tert$-butyl-4-methoxy-phenol.
[f] Taken as $\frac{1}{2}k_{XH/X}$(PhOH) to account for statistical factor of 2, see ref. 55.
[g] Reference 14.
[h] $iAscH^-$ = 5,6-O-isopropylidene ascorbate.
[i] Trolox C = ($\pm$)-6-hydroxy-2,5,7,8-tetramethylchromane-2-carboxylic acid.
[j] DHA = 9,10-dihydroanthracene.
[k] $k_{XH/X}$ are not expected to vary with solvent since α_2^H(C–H) ~ 0.
[l] Estimated using the Marcus cross relation; see references 51 and 91.

Table 1.2 Summary of observed and calculated (eqn (1.8)) organic HAT rate constants.

Entry	Reaction	Solvent $(\beta_2^H)^b$	$K_{XH/Y}$^{c}	k_{obs}^{a}	k_{calc}^{a}	k_{rel}	Ref.
1*	^{t}Bu₃PhO• + TEMPOH	MeCN (0.44)	5×10^7	1.25×10^4	2.9×10^4	2.3	14
2*	^{t}Bu₃PhO• + TEMPOH	DMSO (0.78)	5×10^7	2.7×10^3	7.2×10^3	2.6	14
3*	^{t}Bu₃PhO• + TEMPOH	C₆H₆ (0.14)	3.5×10^8	9.5×10^4	2.2×10^5	4.2	14
4*	^{t}Bu₃PhO• + TEMPOH	CCl₄ (0.05)	3.5×10^8	9.5×10^4	2.2×10^5	6.4	14
5*	^{t}Bu₂MeOPhO• + TEMPOH	MeCN (0.44)	3×10^5	2.67×10^3	3.5×10^3	1.3	14
6*	^{t}Bu₂MeOPhO• + TEMPOH	DMSO (0.78)	3.5×10^4	6.2×10^2	8.3×10^2	1.4	14
7*	^{t}Bu₂MeOPhO• + TEMPOH	C₆H₆ (0.14)	2.5×10^6	1.85×10^4	2.7×10^4	3.1	14
8*	^{t}Bu₃PhO• + iAscH⁻	MeCN (0.44)	2.3×10^8	3.4×10^6	1.3×10^7	3.8	56
9*	^{t}Bu₂MeOPhO• + iAscH⁻	MeCN (0.44)	1.5×10^6	5.3×10^5	1.9×10^6	3.7	56
10	ROO• + ^{t}Bu₂MeOPhOH^d	styrene (0.18)	3.8×10^6	1.1×10^5	4.1×10^4	3.0	57
11	ROO• + BHT^d	styrene (0.18)	4×10^3	1.4×10^4	5.6×10^3	2.8	57
12	ROO• + 2,6-^{t}Bu₂PhOH^d	styrene (0.18)	2×10^2	3.1×10^3	1.4×10^3	2.5	57
13	ROO• + TocOH^e	styrene (0.18)	4.6×10^5	3.2×10^6	1.8×10^6	2.1	57
14*	^{t}BuOO• + ^{t}Bu₂MeOPhOH	alkane^e (0)	2.4×10^5	1.1×10^5	8.8×10^4	1.3	57
15*	^{t}BuOO• + BHT	alkane^e (0)	2.5×10^3	2.4×10^4	1.2×10^4	2.0	57
16	^{t}BuOO• + TocOH	alkane^e (0)	4×10^5	2.6×10^6	3.8×10^6	1.5	57
17	α-Toc• + ^{t}BuOOH	ethanol (0.44)	1.7×10^{-6}	4.1×10^{-1}	3.8×10^{-1}	1.2	58
18	^{t}Bu₃PhO• + tetralin-OOH	PhCl (0.11)	3.9×10^{-4}	3.4×10^{-1}	2×10^0	7.4	59
19	^{t}Bu₃PhO• + PhOH	hexane (0)	3.1×10^{-7}	5.7×10^0	1.9×10^1	3.0	72
20	^{t}BuOO• + 1-naphthol	isopentane (0)	1.2×10^1	1.5×10^5	5.3×10^5	2.0	60
21	^{t}BuOO• + 2-naphthol	isopentane (0)	1.3×10^{-2}	3.1×10^4	1.7×10^4	1.8	61
22	^{t}BuOO• + PhOH	heptane (0)	5.5×10^{-4}	3×10^3	2.8×10^3	1.1	62
23*	^{t}BuO• + PhOH	DTBP:C₆H₆^f	3×10^{12}	3.3×10^8	5.9×10^9	16	63
24*	PhO• + TocOH	2:1 DTBP:MeCN	9×10^9	3.2×10^8	1.2×10^9	3.3	64
25*	PhO• + TocOH	3:1 DTBP:C₆H₆	9×10^9	1.1×10^9	1.2×10^9	1.0	64
26	PhO• + Trolox C	water (0.38)	1×10^8	4.1×10^8	9.7×10^8	2.4	65
27	^{t}Bu₃PhO• + PhOH	PhCl (0.09)	1.6×10^{-7}	$<8 \times 10^{0.0}$ ^{g}	6.6×10^0	1.3	66
28	Trolox C radical + AscH⁻	water (0.38)	4.7×10^3	1.4×10^7 ^{h}	2.5×10^7	1.8	67
29				8.3×10^6 ^{h}		3.0	65
30	tyrosyl radical + AscH⁻	water (0.38)	4×10^{10}	4.4×10^8	7.0×10^9	16	68

Table 1.2 (*Continued*)

Entry	Reaction	Solvent $(\beta_2^H)^b$	$K_{XH/Y}{}^c$	$k_{obs}{}^a$	$k_{calc}{}^a$	k_{rel}	Ref.
31	tyrosyl radical + Trolox C	water (0.38)	9.3×10^6	3.1×10^8	4.5×10^8	1.5	68
32	PhO$^\bullet$ + DHA	PhCl (0.09)	2.3×10^8	$<1.1 \times 10^2$ ^{g}	2.1×10^2	1.7	66
33*	^{t}Bu$_3$PhO$^\bullet$ + DHA	MeCN (0.44)	3.0×10^1	1.8×10^{-3}	3.3×10^{-4}	3.1	14
34*	^{t}BuO$^\bullet$ + DHA	DTBP:C$_6$H$_6$	9×10^{21}	9.5×10^6	7.9×10^5	15	63
35*	^{t}BuO$^\bullet$ + toluene	DTBP:C$_6$H$_6$	3×10^{12}	2.3×10^5	3.4×10^5	2.5	63
36	^{t}BuOO$^\bullet$ + toluene	toluene (0.14)	2.6×10^{-4}	1×10^{-2}	1.5×10^{-3}	9.4	62

*Indicates that both $k_{XH/X}$ and $k_{YH/Y}$ are known in the given solvent.
^{a}k in M^{-1} s^{-1} at 298 K. k_{rel} is defined as k_{obs}/k_{calc} or k_{calc}/k_{obs}, whichever is greater than 1.
^{b}β_2^H values from ref. 36.
^{c}$K_{XH/Y}$ in organic solvents corrected using the Abraham model, except for *. $K_{XH/Y}$ in water from thermochemical cycles.
^{d}ROO$^\bullet$ = peroxylpolystyryl.
^{e}Alkane = decane or cyclohexane.
^{f}DTBP = ^{t}BuO^{t}Bu.
^{g}Rate constant measured at 333 K.
^{h}Independent determinations of the rate constant for Trolox radical + AscH$^-$ give slightly different values.

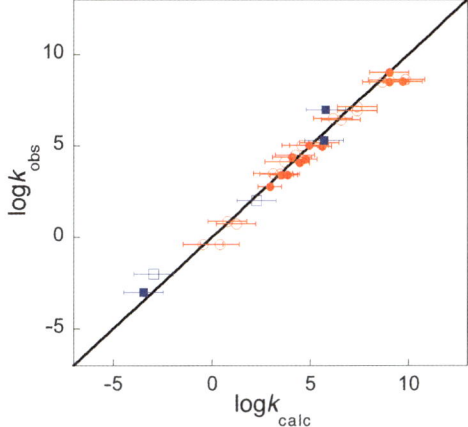

Figure 1.2   Comparison between HAT rate constants measured experimentally (k_{obs})
vs. those determined from the Marcus cross relation (eqn (1.8), k_{calc}) for
reactions where $k_{\mathrm{XH/Y}}$, $k_{\mathrm{XH/X}}$ and $k_{\mathrm{YH/Y}}$ have been measured in the same
solvent. The line indicates perfect agreement. The closed symbols are
reactions where all of the inputs for the cross relation are known (marked
with * in Table 1.2). The reactions (Table 1.2 with *) involve oxyl
radicals + O–H bonds (red ● or ○) or C–H bonds (blue ■ or □).

The agreement between the experimental HAT rate constants and those
predicted using the Marcus cross relation (eqn (1.8)) is remarkably close, as
shown in Figure 1.2. The diagonal line in the Figure represents perfect agreement. The average deviation is a factor of 4.4, and for 14 of the 17 reactions the
cross rate constants ($k_{\mathrm{XH/Y}}$) are predicted to within a factor of 4. These reactions span 10^{12} in $k_{\mathrm{XH/Y}}$, 10^{18} in $K_{\mathrm{XH/Y}}$, and 10^{18} in self-exchange rates. The
implications of this remarkable result are discussed below.

1.3.3   Solvent Effects on HAT Rate and Equilibrium Constants

In some instances self-exchange rate constants are not available in the solvent
of interest. A solvent correction is necessary in these cases because, as shown by
Ingold and Litwinienko, rate constants for H˙ abstraction from O–H or N–H
bonds in organic molecules can vary by more than 10^2 between strongly and
weakly H-bond accepting solvents.[33,69] This is because HAT does not occur
with substrates that are hydrogen bonded to solvent, X–H · · · solvent; the
hydrogen bond to solvent must be cleaved prior to reaction. The Ingold kinetic
solvent effect model (KSE) quantitatively accounts for this effect, using the
empirical Abraham $\alpha_2{}^{\mathrm{H}}$ and $\beta_2{}^{\mathrm{H}}$ parameters to estimate the strength of the
X–H · · · solvent hydrogen bond.[34–36] The model (eqn (1.13)) gives the HAT
rate constant in solvent S (k^{S}) in terms of the rate constant for the same reaction
in a non-hydrogen bonding solvent, such as an alkane (k^0). Table 1.3 shows
how the self-exchange rate constants (and BDFEs, see below) vary with solvent

Table 1.3 Self exchange rate constants and bond dissociation free energies
(BDFEs) for $^{t}Bu_{3}PhO(^{\bullet}/H)$ in different solvents.[a]

Solvent	$k_{XH/X}$	BDFE
Hexane	140 ± 25	76.0
CCl$_4$	130 ± 20	76.7
Benzene	95 ± 14	76.9
MeCN	20 ± 3	77.8
DMSO	8 ± 2	79.2

[a]$k_{XH/X}$ in M^{-1} s^{-1} and BDFE in kcal mol^{-1}. See reference 14.

for 2,4,6-tri-*tert*-butylphenol. This model can also be used to convert rate constants between two solvents without knowing k^0.

$$\log(k^S) = \log(k^0) - 8.3\alpha_2^{H}(XH)\beta_2^{H}(S) \qquad (1.13)$$

The free energy of an HAT reaction also varies with the solvent. ΔG°_{HAT} for $X + H-Y$ is the difference in the BDFEs for X–H and Y–H. The variation of these BDFEs with solvent can be estimated using the same Abraham parameters. As described in more detail elsewhere,[14,39] BDFE(X–H) in solvent S is related to the gas-phase BDFE(X–H) by $\Delta G^{\circ}_{solvation}(H^{\bullet})$ and the difference in the free energies of solution of XH and X, which is primarily due to differences in hydrogen bonding.[70] In polar aprotic solvents that act primarily as hydrogen bond acceptors, the difference in solvation is essentially the strength of the XH $\cdots$ solvent hydrogen bond, which is given by Abraham parameters (eqn (1.14)). In protic solvents, H-bonding between the radical and solvent ($X^{\bullet} \cdots H-S$) also needs to be included. This model gives BDFEs that are in good agreement with values from other methods, such as thermochemical square schemes.[14,39]

$$\Delta G^{\circ}_{solv}(HX) - \Delta G^{\circ}_{solv}(X^{\bullet}) \cong \Delta G^{\circ}_{H\text{-bond}}$$
$$= -10.02\alpha_2^{H}(XH)\beta_2^{H}(S) - 1.492 \qquad (1.14)$$

1.3.4 A Test Case: Reactions of Bulky Phenoxyl Radicals with TEMPOH

As a means to test the combined cross relation/KSE/Abraham model, we examined the reactions of bulky phenoxyl radicals with the hydroxylamine TEMPOH, using cross and self-exchange rate constants in CCl$_4$, C$_6$H$_6$, MeCN and DMSO.[14] The cross reaction is shown in equation 1.15. Self-exchange rate constants for TEMPO($^{\bullet}/H$) were determined using the pseudo-self exchange reaction of TEMPOH and the stable radical 4-oxo-TEMPO (eqn (1.11)).[14,52,53]

$$R = {}^{t}Bu, OMe \tag{1.15}$$

$k_{XH/X}$ values for ${}^{t}Bu_3PhO({}^{\bullet}/H)$ in these solvents were similarly determined from the pseudo-self exchange reaction of isolated ${}^{t}Bu_3PhO^{\bullet}$[71] with 2,6-di-*tert*-butyl-4-methyl-phenol (BHT, ${}^{t}Bu_2MePhOH$), as described elsewhere.[72] Starting with the $k_{XH/X}$ in the least polar solvent, the ${}^{t}Bu_3PhO({}^{\bullet}/H)$ and TEMPO$({}^{\bullet}/H)$ self-exchange rate constants in the various other solvents were calculated using eqn (1.13). The predicted rate constants are within error of the measured values for all but one of the reactions; the calculated value for ${}^{t}Bu_3PhO({}^{\bullet}/H)$ in DMSO deviates by a factor of two, slightly outside the error limits.[14] Thus, as expected,[33] the KSE model holds for these self-exchange reactions.

Using $k_{XH/X}$, $k_{YH/Y}$, and $K_{XH/Y}$ measured in the same solvent, the Marcus cross relation predicts cross rate constants that are all in very good agreement, within a factor of 6.5, of the measured values. To test the combined cross relation/KSE model, cross rate constants in polar solvents have also been calculated using rate constants from less polar media adjusted using the KSE model, as described above. These calculated cross reaction rate constants are again in very good agreement with those directly measured, within a factor of 4.5. This agreement is not surprising since the self-exchange rate constants predicted by the KSE model are close to the measured values. This exercise validates the interchangeable use of KSE-adjusted or directly measured self-exchange rate constants in the cross relation. In the next section, we use this conclusion to test the cross relation with a larger dataset.

1.3.5 Tests of the Cross Relation using KSE-Corrected Self-Exchange Rate Constants

Bringing together the issues described above gives a combined cross relation/kinetic solvent effect (KSE) model. Self-exchange rate constants $k_{XH/X}$ and equilibrium constants $K_{XH/Y}$ measured in different solvents can be used, correcting them to the solvent of interest using the KSE model (eqn (1.13)) or eqn (1.14). These solvent-corrected values are inserted into the cross relation (eqn (1.8) and (1.7)) to give a predicted cross rate constant. This model has been tested for 19 organic reactions in addition to the 17 where all of the inputs were measured in the same solvent (Section 1.3.2, above). These are listed in Table 1.2.

Very good agreement is found between the measured rate constants and those predicted by the cross relation/KSE model, as illustrated in Figure 1.2. The average deviation between k_{calc} and k_{obs} for the collective data set is 3.8. Thirty of the 36 predicted $k_{XH/Y}$ are within a factor of 5 of the experimentally observed rate constants. One notable feature is that the model allows for calculation of rate constants in protic media, such as aqueous reactions of biologically interesting molecules such as ascorbate, even though we know of no reports of HAT self-exchange rate constants in such solvents.

The accuracy of the cross relation/KSE model for a highly diverse set of HAT reactions is remarkable. The agreement even as good as, or better than, is found for application of the cross relation to electron transfer.[73] In all other cases that we are aware of, rate/driving-force relationships of this kind only hold within a set of similar reactions, such as hydride transfers among NADH-analogues.[22] The reactions in Table 1.2 are not similar: they involve O–H and/or C–H bonds, and are done in solvents from alkanes to water. As shown in the next section, the cross relation holds fairly well for HAT reactions of transition metal complexes as well, further increasing the diversity of the reaction set. Various implications of the success of the cross relation/KSE model – and its limitations – will be discussed in Section 1.5 below. We emphasize here that the close agreement validates the conceptual model. HAT occurs from non-hydrogen bonded X–H substrates and its rate is well determined by the combination of the free energy of reaction and the intrinsic barriers, which can be independently determined from self-exchange reactions.

1.4   Predicting HAT Rate Constants for Transition Metal Complexes

Transition metal compounds, complexes, heterogeneous catalysts, and enzyme active sites are involved in a wide variety of important HAT reactions. Among the earliest indications of this mechanism were for organic oxidations by permanganate and chromium(VI) compounds, dating back many decades.[74] However, the generality of this reaction chemistry has only come to light more recently, perhaps starting with the discovery of HAT reactivity of the ferryl (Fe=O) centers in cytochromes P450.[75] HAT is now recognized as probably the most important mechanism for the oxidation of C–H and O–H bonds by transition metal centers. Examples include the use of permanganate and other oxometal reagents in organic syntheses,[76] selective alkane oxidations over oxide surfaces,[77] and many enzymatic processes including aliphatic C–H hydroxylation by methane monooxygenase and other non-heme iron enzymes,[78] fatty acid oxidation by the iron-hydroxide in lipoxygenase,[79] and ascorbate recycling by cytochrome b_{561}.[80] Thus, there is a great deal of interest in understanding the factors that influence the HAT reactivity of transition metal complexes.

The HAT reactions of transition metal complexes described here all involve redox change at the metal coupled to proton transfer to/from a ligand (eqn (1.16)). One example is shown in equation (1.17), in which oxidation of an Os(III) center is coupled to deprotonation of an aniline to an anilide ligand. There is also an extensive literature of HAT reactions of metal-hydride species, $L_nM–H$, which in some ways more closely resemble organic HAT reactions.[81,82]

$$L_nM^{z+}(X:) + H-Y \rightarrow L_nM^{(z-1)+}(XH) + Y \qquad (1.16)$$

$$(1.17)$$

In equation 1.16, there are various ligands X:/XH that can accept/donate the proton. These include the oxo group in the ferryl active sites in heme and non-heme oxygenase enzymes, the hydroxide at the active site in lipoxygenase, the aniline in equation (1.17), and an imidazolate, as in equation (1.18). As these examples illustrate, some metal-mediated HAT reactions involve a formal separation of the e^- and H^+ that make up the transferred $H^{\bullet}$. In the case of metal-imidazole complexes the proton and electron are 3 bonds or ~ 4 Å separated. Despite this separation, we refer to all of these reactions as HAT. Recent interest into the intimate details of HAT reactions has led to new thinking and new definitions.[10–13] Still, we prefer the simplest definition of HAT, as reactions where H^+ and e^- are transferred from one donor to one acceptor in a single kinetic step (eqn (1.1)). The definitions and abbreviations of the organic ligands used in the transition metal HAT systems described below are given in ref. 54.

$$(1.18)$$

Our initial interest in applying the cross relation to HAT grew out of the limitations of the Bell–Evans–Polanyi (BEP) equation discussed above. This equation holds within a set of similar reactions, but with the expansion of HAT reactions to include transition metal reactions it was not clear what made reagents "similar." It was not evident why different classes of reactions fall on different correlation lines (defined by the parameters α and β, see above). For example, it has long been known that, at the same driving force, $H^{\bullet}$ abstraction from O–H bonds is substantially faster than from C–H bonds. Transition metal reagents show the same kinetic pattern, O–H faster than C–H.[5] Clearly, even for the simplest HAT reactions, there is more at play than simply driving force and/or bond strengths. Additional factors are important for transition metal systems, as was revealed when the rates for roughly isoergic HAT reactions of isostructural cobalt- and iron-(tris(bi-imidazoline) complexes differed by many orders of magnitude (see below).[15,47,83]

We have found that the Marcus theory/cross relation approach provides a valuable conceptual framework to address these issues, as well as being a predictive tool. The following sections present a few examples of transition metal HAT systems that have been studied in detail. These build on our first tests of the applicability of the cross relation, presented in 2004, which were mostly reactions of bipyrimidine- and biimidazoline-ligated iron coordination complexes (the top entries in Table 1.3).[15] We then summarize all of the systems that have been analyzed, including updates of the original examples. These and following sections discuss the overall applicability of the cross relation to transition metal HAT reactions and some of the challenges associated with this analysis.

1.4.1 Applying the Cross Relation as a Function of Temperature; the Importance of Using Free Energies

The reaction of iron(II) tris-bi(tetrahydro)pyrimidine ($Fe^{II}H_2bip$) with TEMPO (eqn (1.19)) is a typical transition metal mediated HAT reaction. It is also a very unusual HAT reaction in that it occurs *faster* at lower temperatures ($\Delta H^{\ddagger} = -2.7 \pm 0.4$ kcal mol^{-1}).[53]

$$\tag{1.19}$$

To better understand this unusual effect, all of the inputs of the cross relation were measured between 277 and 328K, allowing application of the cross relation as a function of temperature. These values are illustrated in the combined van't Hoff/Eyring plot in Figure 1.3. There is excellent agreement between the measured cross rate constants (blue dots) and the rate constants calculated with the cross relation, indicated by the red line. *The cross relation quantitatively predicts the negative temperature dependence of the rate constant.* The cross relation also provides an understanding of this unusual effect, *i.e.* that it results from the strong temperature dependence of the equilibrium constant. The reaction proceeds faster at low temperatures because it is significantly more downhill at lower temperatures.

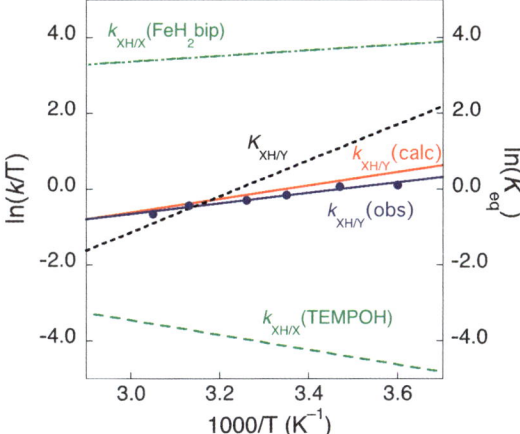

Figure 1.3 Combined Eyring and van't Hoff plot for reaction (19): self-exchange rate constants $k_{XH/X}(FeH_2bip)$ and $k_{XH/X}(TEMPOH)$ (left axis); $K_{HX/Y}$ (right axis), and the rate constants measured ($k_{XH/Y}(obs)$, blue ●) and calculated from the cross relation ($k_{XH/Y}(calc)$ red line) (left axis).[84]

The large temperature dependence of the equilibrium constant for reaction (1.19) was surprising because of the large magnitude of the ground-state entropy change: $\Delta S^\circ = -30 \pm 2 \, cal \, K^{-1} \, mol^{-1}$.[41,42] HAT reactions typically have $|\Delta S^\circ| \sim 0$ because there is no change in charge for a HAT reaction, and usually little change in size. Detailed studies of this and related reactions indicate that this is primarily a change in vibrational entropy (related to the large entropy changes in high-spin/low-spin equilibria).[41,42] The excellent agreement in Figure 1.3 requires that the analysis use the free energy of reaction; if the enthalpy were used the calculated rate constants would deviate by a factor of $\sim 10^3$. These results show that solution HAT reactions, and PCET in general, are best understood using free energies, such as bond dissociation free energies (BDFEs), rather than enthalpies and BDEs as is done in Bell–Evans–Polanyi correlations. In fact, in 1938 Evans and Polanyi derived their equation in terms of free energies, then restated it in terms of enthalpy explicitly assuming that entropic effects were small.[24] Thus, the common use of enthalpies and BDEs has an implicit assumption that $\Delta S^\circ \cong 0$ and $\Delta H^\circ \cong \Delta G^\circ$, an assumption that is probably valid for the large majority of *organic* HAT reactions.[41,42] However, it is more appropriate to use free energies, as is done in the Marcus model, and to identify the Bell–Evans–Polanyi correlation as a linear free energy relationship (LFER) that is a limiting case of the Marcus equation.

1.4.2 Applying the Cross Relation to Oxidations by [Ru^{IV}(O)(bpy)$_2$(py)]$^{2+}$

[Ru^{IV}(O)(bpy)$_2$(py)]$^{2+}$ oxidizes a wide range of organic compounds, as extensively developed by Meyer and coworkers.[3] Our re-examination of the oxidation of alkylaromatic compounds in MeCN showed that these reactions proceed by initial HAT.[29] The ruthenium oxo/hydroxo self-exchange rate constant of 7.6×10^4 M^{-1} s^{-1} was estimated from ^{1}H NMR line broadening measurements, which were challenging because of the concurrent disproportionation of the hydroxide complex Ru^{III}(OH)(bpy)$_2$(py)$^{2+}$. The BDFE for Ru^{III}(OH)(bpy)$_2$(py)$^{2+}$ → Ru^{IV}(O)(bpy)$_2$(py)$^{2+}$ + H$^{\bullet}$ is known in water,[29,84] but not in acetonitrile. The Abraham model can be used to adjust BDFEs between solvents, as noted above, but this can only be done approximately for transition metal species since the relevant α_2^H and β_2^H parameters are not known. With the assumptions that α_2^H(RuO–H) is the same as for H$_2$O or ROH (0.35) and that β_2^H(Ru=O) is similar to that for a ketone (0.5), the [Ru^{III}(O–H)(bpy)$_2$(py)]$^{2+}$ BDE in MeCN is 82.5 ± 1.5 kcal mol^{-1}. The relatively large error bar reflects the uncertainty in the assumptions. Using this value, and the known organic $k_{XH/X}$ rate constants (Table 1.1), the cross relation provides reasonably good predictions. For [Ru^{IV}(O)(bpy)$_2$(py)]$^{2+}$ + toluene, for example, the calculated value of 6.0×10^{-2} M^{-1} s^{-1} is within a factor of 9 of that measured, 6.4×10^{-3} M^{-1} s^{-1}. This system illustrates, however, that it is more difficult to account for kinetic solvent effects in metal-mediated HAT reactions than for organic reagents, and that this is one of the issues that limit the accuracy of the cross relation analysis in these cases.

1.4.3   Precursor and Successor Complexes for HAT

The Marcus theory model is derived for unimolecular electron transfer. It is applied to bimolecular reactions by assuming that the reactants weakly associate in a 'precursor complex' within which ET occurs to give the successor complex.[16–18] The cross relation analyses above have implicitly adopted this same model, but HAT precursor complexes are quite different then ET ones. This is because proton transfer occurs only over very short distances, so HAT precursor complexes have distinct conformations, rather than the weakly interacting encounter complexes of ET. In this way, HAT resembles proton transfer and inner-sphere electron transfer.[85,86] Including the equilibria for precursor and successor complex formation expands equation (1.1) into equation (1.20).

$$\text{X–H} + \text{Y} \underset{}{\overset{K_p}{\rightleftharpoons}} \text{X–H---Y} \underset{}{\overset{K_{HAT}}{\rightleftharpoons}} \text{X---H–Y} \underset{}{\overset{K_{S-1}}{\rightleftharpoons}} \text{X} + \text{H–Y} \qquad (1.20)$$

One case where the importance of HAT precursor and successor complexes is clear is the reaction of [Co(Hbim)(H$_2$bim)$_2$]$^{2+}$ (abbreviated to Co(Hbim)) with TEMPOH (eqn (1.21)).[87] Reaction (1.22) shows kinetic saturation behavior in the forward direction, indicating pre-equilibrium formation of a Co(Hbim)$\cdots$TEMPOH precursor complex with $K_P = 61.3 \pm 0.8$ M^{-1} ($\Delta G_P^{\circ} = -2.44 \pm 0.05$ kcal mol^{-1}). In the reverse direction, saturation was not observed but kinetic models indicate $\Delta G_S^{\circ} = 0.24 \pm 0.83$ kcal mol^{-1}. Taking these values into account, the unimolecular HAT step in reaction (1.21) has $\Delta G_{21}°(\text{HAT}) = -0.3 \pm 0.8$ kcal mol^{-1}, while the net reaction has $\Delta G°_{21} = -3.0 \pm 0.4$ kcal mol^{-1}. This difference of -2.7 ± 1.0 kcal mol^{-1} corresponds to a difference of an order of magnitude in the rate constant predicted by the cross relation.

$$(1.21)$$

A more complete description of HAT reactions would also include the presence of hydrogen-bonded solvent molecules, which (as described in Section 1.3 above) must dissociate prior to formation of the precursor complex (eqn (1.22); formation of the successor complex is similar). A detailed model including these equilibria is presented in reference 14 and its Supporting Information.

$$(1.22)$$

For the organic reactions discussed above, the magnitude of the equilibria can be fairly well predicted using Abraham's hydrogen bonding model, although some parameters have to be estimated, such as the hydrogen bonding basicities (β_2^{H}) of the organic radicals. Fortunately, the effects of these equilibria prior to the HAT step are typically small. The effects are larger for reactions in protic solvents, and for reactions where the reactants and products have very different hydrogen-bonding properties, such as reactions of oxyl radicals with C–H bonds (RO$^{\bullet}$ + R$'$H $\rightarrow$ ROH + R$'^{\bullet}$).[14] Even in these cases, including H-bond equilibria in the analysis usually causes shifts that are within the uncertainty of the calculated cross rate constant (typically *ca.* an order of magnitude), so this can be neglected in most analyses of organic reactions. The situation is less clear for HAT reactions of transition metal complexes because the H-bonding properties of $\text{L}_n\text{M}^{z+}(\text{X:})$ can be very different than those of Y$^{\bullet}$ (eqn (1.16)) and the Abraham parameters are not available. As shown by the cobalt example above, this effect can be significant.

1.4.4 Applying the Cross Relation for Transition Metal HAT

Since the 2004 report applying the cross relation to transition metal HAT reactions, we have examined number of new systems, including $(acac)_2Ru(py\text{-}imH)$,[48] $(bpy)_2V(O)(OH)$,[51] and $TpOs(NH_2Ph)Cl_2$ complexes[49] (Scheme 1.1 and eqn (1.17); ligand abbreviations given in ref. 54). As discussed above, we have discovered that entropic effects can be quite important and free energies of reaction should be used.[42] We also now understand the importance of solvent effects on both self-exchange and equilibrium constants (Section 1.3.3 above). For instance, the newly measured $k(^tBu_3PhO^{\bullet}/H)$ self-exchange rate in MeCN[14] is more than a factor of 10 smaller than the previously reported value in CCl_4 that was used in our original analysis.

For all of these reasons, it is appropriate to update and extend the tests of the cross relation for transition metal HAT reactions. The results are given in Table 1.4 and shown in Figure 1.4. The analyses use the BDFEs in Table 1.1 to calculate the $K_{XH/Y}$ values (instead of the originally used BDEs),[15] except for the cases for which the equilibrium constant could be measured directly (entries 1, 8, 9, and 16 in Table 1.4). While there is somewhat more scatter than for the organic reactions, there is generally good agreement between the observed and calculated cross rate constants. Only four of the 29 calculated rate constants deviate from the fit line by more than two orders of magnitude. These include the reactions of $TpOs^{III}(NH_2Ph)Cl_2$ with aminoxyl radicals, and reactions of $V^V(O)_2(bpy)_2^+$ or $Ru^{IV}(O)(bpy)_2(py)^{2+}$ with xanthene, which are discussed below. Excluding these cases, the average deviation is 15 and about half of the values are predicted to better than a factor of ten. The overall agreement to within 1–2 orders of magnitude is remarkable for a very disparate set of metal complexes, including first, second and third row transition metal ions, with oxo, imidazole and aniline ligands, and with various spin states.

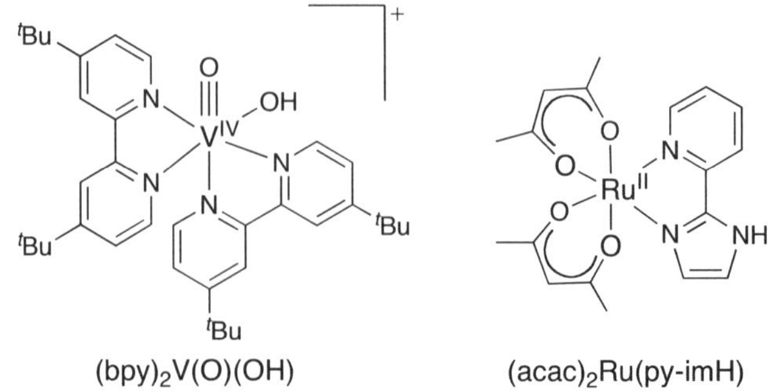

(bpy)$_2$V(O)(OH)   (acac)$_2$Ru(py-imH)

Scheme 1.1   Additional transition metal systems for investigations of HAT reactions.

Table 1.4 Summary of observed and calculated (eqn (1.7)) rate constants for transition metal PCET reactions.[a]

Entry	Reaction	K_{XY}	k_{obs}[a]	k_{calc}[a]	k_{rel}[b]	Ref.
1	$Fe^{III}Hbim^{2+}$ + TEMPOH	5.0×10^3	3.1×10^3	3.9×10^3	1.2	15,41
2	$Fe^{III}Hbim^{2+}$ + Et_2NOH	6.3×10^{-1}	1.1×10^1	5.6×10^1	9.2	15
3	$Fe^{III}Hbim^{2+}$ + H_2Q	2.5×10^{-5}	2.8×10^1	4.0×10^1	1.4	15
4	$Fe^{II}H_2bim^{2+}$ + $^tBu_3PhO^•$	1.7×10^4	6.8×10^5	1.4×10^4	50	15
5	$Fe^{II}H_2bim^{2+}$ + $PhCH_2$	1.1×10^{11}	3.0×10^3	2.0×10^4	6.6	15
6	$Fe^{II}Hbim^{2+}$ + xanthene	1.1×10^{-1}	1.5×10^{-3}	7.8×10^{-5}	14	88
7	$Fe^{III}Hbip^{2+}$ + TEMPOH	5.9×10^{-1}	2.6×10^2	7.2×10^1	3.7	15,41
8	$Fe^{II}H_2bip^{2+}$ + TEMPO	1.7×10^0	1.5×10^2	1.2×10^2	1.2	15,41
9	$Fe^{II}H_2bip^{2+}$ + $^tBu_2NO^•$	5.7×10^{-2}	6.3×10^0	7.6×10^1	6.5	15
10	$Fe^{II}H_2bip^{2+}$ + $^tBu_3PhO^•$	2.0×10^8	1×10^7	8.7×10^5	11.6	15
11	$Ru^{II}(acac)_2(pyimH)$ + TEMPO	1.8×10^3	1.4×10^3	4.3×10^4	31	48
12	$TpOs^{III}(NH_2Ph)Cl_2$ + TEMPO	4.6×10^3	4×10^{-2}	4.8×10^0	120	49
13	$TpOs^{III}(NH_2Ph)Cl_2$ + $^tBu_2NO$	1.9×10^3	2×10^{-2}	3.3×10^0	300	49
14	$V^V(O)_2(bpy)_2^+$ + TEMPOH	1.0×10^3	1×10^{-1}	5.0×10^0	50	51
15	$V^V(O)_2(bpy)_2^+$ + xanthene	1.7×10^{-2}	3.8×10^{-5}	5.8×10^{-7}	370	51
16	$V^V(O)_2(bpy)_2^+$ + $^tBu_2OMePhOH$	1.4×10^{-3}	1.4×10^{-3}	1.2×10^{-2}	8.6	51
17	$V^V(O)_2(bpy)_2^+$ + DHA^c	2.5×10^{-4}	2×10^{-7}	1.3×10^{-8}	25	51
18	$V^V(O)_2(bpy)_2^+$ + H_2Q^d	3.7×10^{-6}	1×10^{-2}	4.0×10^{-2}	4.0	51
19	$Ru^{IV}(O)(bpy)_2py^{2+}$ + DHA^c	1×10^5	1.6×10^1	5.4×10^{-1}	30	29b
20	$Ru^{IV}(O)(bpy)_2py^{2+}$ + xanthene	1×10^7	5.8×10^2	5.6×10^0	100	29b
21	$Ru^{IV}(O)(bpy)_2py^{2+}$ + toluene	8×10^{-4}	6.4×10^{-3}	6.0×10^{-2}	9	29b
22	$Ru^{IV}(O)(bpy)_2py^{2+}$ + fluorene	1×10^4	5.5×10^0	1.8×10^{-1}	30	29b
23	$Mn_2O_2(phen)_4^{3+}$ + xanthene	1.7×10^1	$6.7 \times 10^{-3\,e}$	2.6×10^{-3}	2.6	89
24	$Mn_2O_2(phen)_4^{3+}$ + fluorene	2.0×10^{-2}	$6.2 \times 10^{-4\,f}$	8.6×10^{-5}	7.2	89
25	$Mn_2(O)(OH)(phen)_4^{3+}$ + xanthene	5.9×10^{-3}	$6.4 \times 10^{-3\,e}$	4.0×10^{-4}	16	89
26	$Mn_2(O)(OH)(phen)_4^{3+}$ + fluorene	6.9×10^{-6}	$4.3 \times 10^{-4\,f}$	1.1×10^{-5}	40	89

[a] k in $M^{-1}\,s^{-1}$ in MeCN solvent at 298 K, unless otherwise noted. Ligand abbreviations are given in ref. 54.
[b] k_{rel} is defined as k_{obs}/k_{calc} or k_{calc}/k_{obs}, whichever is greater than 1.
[c] DHA = 9,10-dihydroanthracene.
[d] H_2Q = 1,4-hydroquinone.
[e] Rate constant measured at 292 K.
[f] Rate constant measured at 333 K.

1.4.5 Transition Metal Systems that Deviate from the Cross Relation

A number of reactions of transition metal complexes show significant deviations from the predictions of the cross relation, as noted above. The reaction of $TpOs^{III}(NH_2Ph)Cl_2$ with TEMPO (eqn (1.17)) may deviate because of uncertainties in the $TpOs^{III}(NH_2Ph)Cl_2$ self-exchange rate constant. HAT self-exchange in this system is substantially slower than related electron and proton transfer self-exchanges, and as a result measurements of HAT reactions are plagued by catalysis by trace acid or base.[49] The $TpOs^{III}(NH_2Ph)Cl_2$ system may also deviate due to significant steric crowding. The cross relation implicitly assumes that the transition state for the cross reaction resembles the self-exchange transition states, but if one of the reagents is sterically encumbered its

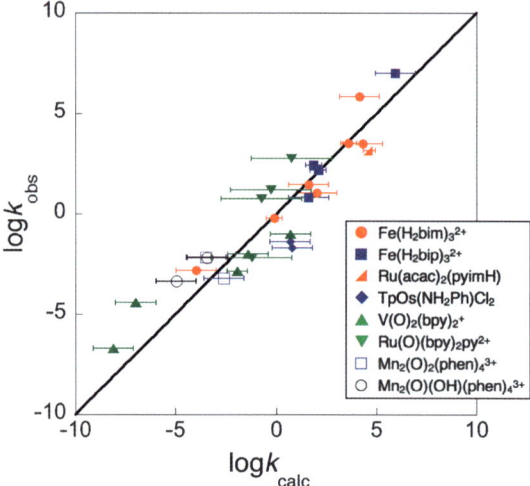

Figure 1.4 Comparison between HAT rate constants measured experimentally (k_{obs})
vs. those determined from the cross relation (eqn (1.8), k_{calc}). The line
indicates perfect agreement. The errors bars are one log unit, except for
$Ru^{IV}(O)(bpy)_2(py)^{2+}$ oxidations, which are 2 log units, reflecting the
uncertainty in its bond strength in MeCN solvent (see text). The correlation coefficient (R^2) for all data is 0.82.

self-exchange process may be inhibited in a way that is not reflected in the cross
reaction (in essence, an effect of the precursor complex). However, this should
lead to calculated cross rate constants that are too small, while the calculated k
for the $TpOs^{III}(NH_2Ph)Cl_2$ reactions are too large.

The reaction of $(acac)_2Ru(py-imH)$ with TEMPO shows a significant
deviation, considering that the estimated errors are small because all of
the inputs for cross relation analysis have been directly measured, including
$K_{XH/Y}$.[48] This reaction shows substantial tunneling, with a hydrogen/deuterium
kinetic isotope effect (KIE) of 23 ± 3 at 298 K. Interestingly, the Ru self-
exchange reaction has a KIE close to 1,[48] but the TEMPO self-exchange
reaction has a large KIE and tunneling is again implicated.[52] The Marcus cross
relation is essentially a classical theory that does not include nuclear tunneling,
so good agreement with experiment is not necessarily expected when tunneling
contributes substantially to HAT rates. (This could also be the case for
$TpOs^{III}(NH_2Ph)Cl_2$, but an isotope effect could not be obtained.[49]) Hydrogen
tunneling has been shown to be integral to many biological processes.[90]

The origin of the deviation from the cross relation for the reactions of
$[V^V(O)_2(bpy)_2]^+$ or $[Ru^{IV}(O)(bpy)_2(py)]^{2+}$ with xanthene is not known. Three
other HAT reactions of xanthene do not show this large deviation (Table 1.4),
so the deviations are not likely to be due to an erroneous xanthene self-
exchange rate constant [taken as that of 9,10-dihydroanthracene[91] (DHA,
Table 1.1)]. Tunneling could be important in reactions of $Ru^{IV}(O)(bpy)_2(py)^{2+}$,
as the reaction with DHA has $k_H/k_D \geq 35$.[29b] Interestingly, the cross relation

appears to hold for the reaction of $Ru^{IV}(O)(bpy)_2(py)^{2+}$ with DHA (Table 1.4), but not for that with xanthene.

These examples highlight that while the cross relation works for many cases, it is a simplified treatment and does not account for a number of other factors that influence transition metal HAT rates. It cannot, for instance, predict when tunneling will be important. Still, the general success of the cross relation, as shown in Figure 1.4, indicates that it captures the largest contributors to HAT reactivity: the free energy of reaction and the intrinsic barrier as measured by self-exchange rate constants. For example, the reactions of $V^V(O)_2(bpy)_2^+$, while they may not all quantitatively follow the cross relation, are all very slow because of a large intrinsic barrier.[51] The $V^V(O)_2(bpy)_2^+/V^{IV}(O)(OH)(bpy)_2^+$ self-exchange rate constant is a million times slower than that of $Ru^{IV}(O)(bpy)_2(py)^{2+}/Ru^{IV}(OH)(bpy)_2(py)^{2+}$, and this is clearly reflected in the cross reaction rates. The origin of this striking difference in intrinsic barriers has been traced to a larger inner-sphere reorganization energy in the vanadium system, due in large part to the sizable changes in the lengths of the strong vanadium–oxygen bonds.[51] Thus, the qualitative Marcus picture of inner-sphere reorganization energies, developed for electron transfer, also holds for HAT. In addition to the intuition that carries over from electron transfer, PCET and HAT also bring the additional issues associated with the proton transfer component. These are presumably responsible for the dramatically lower reactivity of C–H *vs.* O–H bonds,[5] but a consensus on the origin of this difference has not yet been reached.[92]

1.5 Conclusions: Implications and Limitations of the Cross Relation for Hydrogen Atom Transfer Reactions

A kinetic model has been developed for hydrogen atom transfer (HAT) reactions using the Marcus cross relation. The model has been used to predict rate constants for 62 HAT reactions, including both purely organic reactions and reactions involving transition metal complexes. The reactions involve cleavage and formation of C–H, O–H, and N–H bonds, and occur in solvents ranging from alkanes to water. The systems examined include biologically important HAT reactions involving tocopherol, ascorbate, hydroperoxides, transition metal-oxo compounds, and non-heme iron complexes. The hydrogen abstractors can be organic radicals in which the X–H bond forms at the site of high unpaired spin density, or transition metal complexes in which there are no unpaired spins. The transition metal reactions involve redox change at the metal coupled to protonation/deprotonation of a ligand, for instance interconversion of $L_nV^V=O$ with $L_nV^{IV}–OH$. The 62 reactions include data that have been measured by different groups, using different physical techniques, over 50 years. They span more than 10^{17} in rate constant, 10^{28} in equilibrium constant, and 10^{18} in the self-exchange rate constants of the reactants.

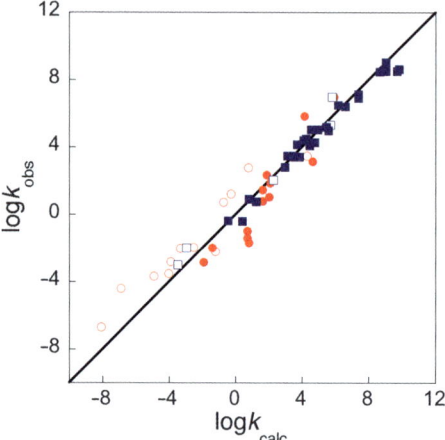

Figure 1.5 Comparison of 63 organic-only and transition metal HAT rate constants
measured experimentally (k_{obs}) *vs.* those determined from the cross relation (eqn (1.8), k_{calc}). The line indicates perfect agreement. The reactions
(Tables 1.2 and 1.4) involve oxyl radicals + O–H bonds (blue ■), oxyl
radicals + C–H bonds (blue □), transition metal complexes + O–H bonds
(red ●) and transition metal complexes + C–H bonds (red ○). Error bars
are omitted for clarity (see Figures 1.2 and 1.4). The correlation coefficient
(R^2) for all data is 0.94.

Over this wide variety of reactions, the cross relation predicts the rate
constants very well, as shown in the combined plot showing all 62 reactions in
Figure 1.5. The correlation coefficient is 0.94 for all data, and rate constants for
46 of the 62 reactions are predicted to within a factor of 9 of the observed rate
constant. Only four of the predicted rate constants deviate by more than a
factor of 10^2.

The success of this model is notable for a number of reasons. In particular, it
is remarkable that the model holds so well for such a wide variety of reactions
and reactants. Linear free energy relationships (LFERs) relating rate constants
with driving force (*e.g.*, Brønsted relationships) are a very useful part of
reaction chemistry, but they are essentially always limited to a set of closely
related compounds and reactions. LFERs such as $\Delta G^{\ddagger} = \alpha \Delta G^{\circ} + \beta$ have parameters (α,β) that are defined only by this relationship. In contrast, the values
that enter into the cross relation, $K_{XH/Y}$, $k_{XH/X}$ and $k_{YH/Y}$, and the parameters
for the KSE model (α_2^{H} and β_2^{H}), are all independently measured and have
independent meaning. *There are no adjustable or fitted parameters in this model.*

This general agreement with the cross relation appears to hold only for HAT
and for outer-sphere electron transfer.[22,41] The other examples of application of
the cross relation, to H$^+$ transfer, H$^-$ transfer, and S$_N$2 reactions, only hold
within a limited subset of similar reactions.[21–23] The critical feature here
appears to be the broad applicability of the additivity postulate for HAT, that
the intrinsic barrier for a reaction is well estimated by the average of the

intrinsic barriers to the self-exchange reactions (eqn (1.4)). This postulate is intuitively reasonable for outer-sphere electron transfer reactions, in which no bonds are made or broken and in which precursor complexes are loosely associated, but it is very surprising that it holds so well for HAT. As pointed out by Sutin in the context of electron transfer,[17] the success of the cross relation is in part a result of its inherent averaging. Still, the cross reactions do not always resemble the average of the self reactions. In RO$^\bullet$ abstractions from C–H bonds, for example, the RO$^\bullet \cdots$H–R transition state does not have a significant hydrogen bond, but in most cases the RO$^\bullet$ + H–OR self-exchange transition state does.

A class of HAT reactions for which the additivity postulate appears not to hold are those with strong "polar effects." In some HAT reactions, as pointed out by Tedder, "...the rate of atom transfer is very dependent on the degree of polarity in the transition state."[93] For instance, Rong *et al.* showed that alkyl radicals abstract H$^\bullet$ faster from thiols than from silanes or stannanes, while the kinetic preference is reversed for perfluoroalkyl radicals.[94] The more electron rich R$^\bullet$ radical preferentially abstracts the electron deficient RS$^{\delta-}$–H$^{\delta+}$ while the electron deficient R$_F^\bullet$ radical reacts faster with R$_3$Sn$^{\delta+}$–H$^{\delta-}$. Such an inversion of reactivity cannot be accounted for by a cross relation treatment, because from the additivity postulate the reactivity of a reagent is not dependent on its partner.

The success of the cross relation for HAT supports the conclusion that these reactions need to be analyzed using free energies. We advocate revising the traditional Bell–Evans–Polanyi (BEP) relation to use ΔG° rather than ΔH°. This will require a change from bond dissociation enthalpies (BDEs) to bond dissociation free energies (BDFEs), which we have tabulated in a recent review.[39] The use of free energies is also consistent with the Ingold kinetic solvent effect (KSE) model and Abraham hydrogen bonding parameters, as these are also based on ΔG°'s.

The agreement with the cross reaction is slightly better for the organic reactions than for those reactions involving transition metal complexes. This is in part because corrections could be made for solvent effects when necessary in the organic case. Ingold's kinetic solvent effect (KSE) analysis and Abraham's hydrogen bonding parameters allow solvent effects to be accounted for when the inputs of the cross relation ($k_{XH/X}$, $k_{YH/Y}$, $K_{XH/Y}$) are not all measured in the same solvents. This combined model is shown to be very successful at predicting HAT rate constants. For transition metal complexes, however, the Abraham parameters are not available and therefore solvent hydrogen bonding effects and precursor complex formation are more difficult to account for.

The Marcus cross relation (and the KSE model) are also valuable as indicators of mechanism. Good agreement between experiment and theory is a strong indication that all of the kinetic components, $k_{XH/Y}$, $k_{XH/X}$, and $k_{YH/Y}$ are for processes that occur by concerted HAT mechanisms. Of course, $K_{XH/Y}$ must refer to the thermodynamics of the overall H$^\bullet$ transfer as well. Similar arguments have been made for the application of the cross relation to ET, and Ingold *et al.* have made this point about the KSE model.[33] For instance, one set of reactions where the KSE model fails was shown to follow a mechanism of sequential proton transfer then electron transfer (PT/ET), rather than HAT. We do not believe, however, that disagreement with the cross relation necessarily means that a non-HAT mechanism is being followed. The reactions in Tables 1.2 and 1.4 for which k_{calc} differs substantially from k_{obs} are still indicated to be HAT, from thermochemical arguments. The observed deviations from the cross relation are likely due to other factors, such as the effects of precursor/successor complexes, hydrogen tunneling, steric effects or non-adiabatic effects.

It should be emphasized that the cross relation is conceptually a very simplified model. It ignores many of the lessons taught by sophisticated treatments of HAT and PCET, such as Hammes–Schiffer's multistate continuum theory.[13] This theory has been applied to many interesting systems, for instance explaining the unusual temperature dependence of a quinol oxidation[95] and the extremely large KIEs in lipoxygenase.[96] The theory uses a non-adiabatic Marcus theory-type approach, treating both the electron transfer and the proton transfer in a quantum mechanical fashion. It therefore includes both electronic and vibrational couplings, summed over vibrational levels and integrated over a range of proton donor–acceptor distances. The cross relation, as applied here, ignores or simplifies all of these factors. We reiterate that the cross relation is viewed as being successful when its prediction within an order of magnitude or two of the observed rate constant. This cross relation analysis is not appropriate to study finer issues of HAT reactions, such as isotope effects.

Most importantly, the success of this Marcus-based model indicates that it captures the primary determinants of the rate of an HAT reaction: the driving force $K_{XH/Y}$ ($\Delta G^\circ_{XH/Y}$), the intrinsic barrier λ, which can be determined from self-exchange rates, and solvent–solute hydrogen bonding. To a first approximation, the electronic structure of the reactants are only important as they influence ΔG° and λ. Notably, there is no evidence for the spin or 'radical character' of the oxidant having a significant influence on the facility of HAT reactions, despite the common intuition to the contrary. As emphasized elsewhere,[15c] measured or estimated self-exchange rate constants are not higher for species with more spin density on the atom that accepts the proton. Diamagnetic compounds such as permanganate and CrO_2Cl_2 are reactive H-atom abstractors despite being diamagnetic.[5,25]

The conceptual picture for HAT developed here places the essentially kinetic information in the intrinsic barriers (self-exchange rate constants). Our understanding of these HAT intrinsic barriers is still limited, but some patterns are emerging. C–H bonds are intrinsically much less reactive than O–H bonds, as shown by the values of $k_{XH/X}$ in Table 1.1. For transition metal reagents, there is evidence that the factors that contribute to electron transfer inner-sphere reorganization energies will also contribute to HAT intrinsic barriers. In the vanadium–oxo system, for instance, large changes in bond lengths of the high frequency V–O bonds are the primary origin of the very slow self-exchange rate constant.[51] Thus the cross relation + KSE model, while a simplification, is a conceptual and predictive tool that can be used to understand a wide range of solution HAT reactions.

References

1. *Hydrogen Transfer Reactions*, ed. J. T. Hynes, J. P. Klinman, H.-H Limback and R. L. Schowen, Wiley-VCH, Weinheim, Germany, 2007.
2. M. Pourbaix, *Atlas d'Equilibres Electrochemiques*, Ganthier-Villars, Paris, 1963.
3. M. H. V. Huynh and T. J. Meyer, *Chem. Rev.*, 2007, **107**, 5004.
4. C. Costentin, *Chem. Rev.*, 2008, **108**, 2145.
5. J. M. Mayer, *Ann. Rev. Phys. Chem.*, 2004, **55**, 363.
6. *Cf.* (a) E. R. Young, J. Rosenthal, J. M. Hodgkiss and D. G. Nocera, *J. Am. Chem. Soc.*, 2009, **131**, 7678; (b) R. I. Cukier and D. G. Nocera, *Annu. Rev. Phys. Chem.*, 1998, **49**, 337.
7. D. Njus and P. M. Kelley, *Biochim. Biophys. Acta*, 1993, **1144**, 235.
8. C. Costentin, D. H. Evans, M. Robert and J.-M. Savéant, *J. Am. Chem. Soc.*, 2005, **127**, 12490.
9. (a) *Free Radicals*, ed. J. K. Kochi, Wiley, New York, 1973; (b) M. J. Perkins, *Free Radical Chemistry*, Ellis Horwood, New York, 1994.
10. (a) J. M. Mayer, D. A. Hrovat, J. L. Thomas and W. T. Borden, *J. Am. Chem. Soc.*, 2002, **124**, 11142; (b) M. Lingwood, J. R. Hammond, D. A. Hrovat, J. M. Mayer and W. T. Borden, *J. Chem. Theory Comput.*, 2006, **2**, 740.
11. R. A. Binstead, M. E. McGuire, A. Dovletoglou, W. K. Seok, L. E. Roecker and T. J. Meyer, *J. Am. Chem. Soc.*, 1992, **114**, 173.
12. O. Tishchenko, D. G. Truhlar, A. Ceulemans and M. T. Nguyen, *J. Am. Chem. Soc.*, 2008, **130**, 7000.
13. S. Hammes-Schiffer and A. V. Soudackov, *J. Phys. Chem. B*, 2008, **112**, 14108.
14. J. J. Warren and J. M. Mayer, *Proc. Natl. Acad. Sci. U. S. A.*, 2010, **107**, 5282.
15. (a) J. P. Roth, J. C. Yoder, T.-J. Won and J. M. Mayer, *Science*, 2004, **294**, 2524. For additional perspective on applying Marcus-type treatments to HAT reactions see: (b) J. M. Mayer, *J. Phys. Chem. Lett.*, 2011, **2**, 1481–1489; (c) J. M. Mayer, *Acc. Chem. Res.*, 2010, **44**, 36–46.
16. R. A. Marcus and N. Sutin, *Biochim. Biophys. Acta*, 1985, **811**, 265.
17. N. Sutin, *Prog. Inorg. Chem.*, 1983, **30**, 441.
18. T. J. Meyer and H. Taube, in *Comprehensive Coordination Chemistry*, ed. G. Wilkinson, Permagon, New York, 1987, vol. 1, p. 331.
19. P. F. Barbara, T. J. Meyer and M. A. Ratner, *J. Phys. Chem.*, 1996, **100**, 13148.
20. (a) R. A. Marcus, *J. Phys. Chem.*, 1968, **72**, 891; (b) A. O. Cohen and R. A. Marcus, *J. Phys. Chem.*, 1968, **72**, 4249; (c) R. A. Marcus, *J. Phys. Chem. A*, 1997, **101**, 4072; (d) R. A. Marcus, *J. Phys. Chem. B*, 2007, **111**, 6643.
21. W. J. Albery, *Ann. Rev. Phys. Chem.*, 1980, **31**, 227.

22. (a) I.-S. H. Lee, E. H. Jeoung and M. M. Kreevoy, *J. Am. Chem. Soc.*, 1997, **119**, 2722; (b) I.-S. H. Lee, K.-H. Chow and M. M. Kreevoy, *J. Am. Chem. Soc.*, 2002, **124**, 7755.

23. (a) A. Pross and S. S. Shaik, *J. Am. Chem. Soc.*, 1982, **104**, 1129; (b) S. S. Shaik and A. Pross, *J. Am. Chem. Soc.*, 1982, **104**, 2708.

24. M. G. Evans and M. Polanyi, *Trans. Faraday Soc.*, 1938, **34**, 11.

25. J. M. Mayer, *Acc. Chem. Res.*, 1998, **31**, 441.

26. S. S. Shaik, H. B. Schlegel and S. Wolfe, *Theoretical Aspects of Physical Organic Chemistry: The SN2 Mechanism*, Wiley, New York, 1992, p. 11.

27. *Cf.* F. A. Carey, *Organic Chemistry*, McGraw-Hill, New York, 5th edn, 2003, p. 170.

28. (a) G. K. Cook and J. M. Mayer, *J. Am. Chem. Soc.*, 1995, **117**, 7139; (b) K. A. Gardner, L. L. Kuehnert and J. M. Mayer, *Inorg. Chem.*, 1997, **36**, 2069.

29. (a) T. Matsuo and J. M. Mayer, *Inorg. Chem.*, 2005, **44**, 2150; (b) J. R. Bryant and J. M. Mayer, *J. Am. Chem. Soc.*, 2003, **125**, 10351; (c) J. R. Bryant, T. Matsuo and J. M. Mayer, *Inorg. Chem.*, 2004, **43**, 1587.

30. See: A. Gunay and K. H. Theopold, *Chem. Rev.*, 2010, **110**, 1060 and, as representative examples: (a) C. R. Goldsmith, R. T. Jonas and T. D. P. Stack, *J. Am. Chem. Soc.*, 2002, **124**, 83; (b) J. Kaizer, E. J. Klinker, N. Y. Oh, J.-U. Rohde, W. J. Song, A. Stubna, J. Kim, E. Münck, W. Nam and L. Que Jr., *J. Am. Chem. Soc.*, 2004, **126**, 472; (c) D. E. Lansky and D. P. Goldberg, *Inorg. Chem.*, 2006, **45**, 5119; (d) C. V. Sastri, J. Lee, K. Oh, Y. J. Lee, J. Lee, T. A. Jackson, K. Ray, H. Hirao, W. Shin, J. A. Halfen, J. Kim, L. Que Jr., S. Shaik and W. Nam, *Proc. Nat. Acad. Sci. U. S. A.*, 2007, **104**, 19181. (e) S. R. Bell and J. T. Groves, *J. Am. Chem. Soc.*, 2009, **131**, 9640.

31. J. M. Mayer, *Acc. Chem. Res.*, 1998, **31**, 441.

32. (a) G. W. Burton and K. U. Ingold, *Acc. Chem. Res.*, 1986, **19**, 194; (b) V. W. Bowry and K. U. Ingold, *Acc. Chem. Res.*, 1999, **32**, 27.

33. G. Litwinienko and K. U. Ingold, *Acc. Chem. Res.*, 2007, **40**, 222.

34. M. H. Abraham, P. L. Grellier, D. V. Prior, R. W. Taft, J. J. Morris, P. J. Taylor, C. Laurence, M. Berthelot, R. M. Doherty, M. J. Kamlet, J.-L. M. Abboud, K. Sraidi and G. Guihéneuf, *J. Am. Chem. Soc.*, 1988, **110**, 8534.

35. M. H. Abraham, P. L. Grellier, D. V. Prior, P. P. Duce, J. J. Morris and P. J. Taylor, *J. Chem. Soc. Perkin Trans.*, 1989, **2** 699.

36. M. H. Abraham, P. L. Grellier, D. V. Prior, J. J. Morris and P. J. Taylor, *J. Chem. Soc., Perkin Trans.*, 1990, **2**, 521.

37. Z is taken as 10^{11} M^{-1} s^{-1} (as is typical (ref. 38)), but even taking Z to be 10^9 M^{-1} s^{-1} does not cause substantial changes in $k_{XH/Y,calc}$ for the HAT reactions discussed here (usually less than a factor of 3).

38. J. H. Espenson, *Chemical Kinetics and Reaction Mechanisms*, McGraw-Hill, New York, 1995, p. 243.

39. J. J. Warren, T. A. Tronic and J. M. Mayer, *Chem. Rev.*, 2010, **110**, 6961.

40. (a) F. G. Bordwell, J.-P. Cheng, G.-Z. Ji, A. V. Satish and X. Zhang, *J. Am. Chem. Soc.*, 1991, **113**, 9790; (b) M. Tilset, in *Electron Transfer in Chemistry*, ed. V. Balzani, Wiley-VCH, New York, 2001, vol. 2, p. 677.

41. E. A. Mader, E. R. Davidson and J. M. Mayer, *J. Am. Chem. Soc.*, 2007, **129**, 5153.
42. E. A. Mader, V. W. Manner, T. F. Markle, A. Wu, J. A. Franz and J. M. Mayer, *J. Am. Chem. Soc.*, 2009, **131**, 4335.
43. (a) Y.-R. Luo, *Comprehensive Handbook of Chemical Bond Energies*, CRC Press, Boca Raton, 2007; (b) Portions are available online: http://books.google.com.
44. J. Sandström, *Dynamic NMR Spectroscopy*, Academic Press, London, 1982.
45. M. R. Arick and S. I. Weissman, *J. Am. Chem. Soc.*, 1968, **90**, 1654.
46. J. P. Roth, S. Lovell and J. M. Mayer, *J. Am. Chem. Soc.*, 2000, **122**, 5486.
47. J. C. Yoder, J. P. Roth, E. M. Gussenhoven, A. S. Larsen and J. M. Mayer, *J. Am. Chem. Soc.*, 2003, **125**, 2629.
48. A. Wu and J. M. Mayer, *J. Am. Chem. Soc.*, 2008, **130**, 14745.
49. J. D. Soper and J. M. Mayer, *J. Am. Chem. Soc.*, 2003, **125**, 12217.
50. R. A. Jackson and D. W. O'Neill, *J. Chem. Soc., Chem. Comm.*, 1969, 1210.
51. C. R. Waidmann, X. Zhou, E. A. Tsai, W. Kaminsky, D. A. Hrovat, W. T. Borden and J. M. Mayer, *J. Am. Chem. Soc.*, 2009, **131**, 4729.
52. A. Wu, E. A. Mader, A. Datta, D. A. Hrovat, W. T. Borden and J. M. Mayer, *J. Am. Chem. Soc.*, 2009, **131**, 11985.
53. E. A. Mader, A. S. Larsen and J. M. Mayer, *J. Am. Chem. Soc.*, 2004, **126**, 8066.
54. The abbreviations for ligands in the transition metal systems described in this chapter are as follows: bpy = 2,2′-bipyridine, ^{t}Bu$_2$bpy = 4,4′-di-*tert*-butyl-2,2′-bipyridine py = pyridine, H$_2$bip = 2,2′-bi-1,4,5,6-tetrahydro-pyrimidine, H$_2$bim = 2,2′-bi-2-imidazoline, phen = phenanthroline, acac = 2,4-pentanedionato, pyimH = 2-(2′-pyridyl)imidazole, Tp = hydrotris(pyrazyl)-borate.
55. The value for the hydroquinone (H$_2$Q) self-exchange rate constant here is different from the value that we used in our original report[15] (from E. B. Zavelovich, A. I. Prokof'ev, *Chem. Phys. Lett.*, 1974, **29**, 212). The previous value was for a reaction of *ortho*-hydroquinones, which have been shown to have reactivity distinct from *para*-hydroquinones (*cf.*, M. C. Foti, R. C. Barclay, K. U. Ingold, *J. Am. Chem. Soc.*, 2002, **124**, 12881). Instead, we now approximate $k_{XH/X}(H_2Q) \sim \frac{1}{2} k_{XH/X}(PhOH)$, which are structurally very similar. The statistical factor of 2 accounts for the fact that there are twice as many reactive H$^{\bullet}$ in H2Q.
56. J. J. Warren and J. M. Mayer, *J. Am. Chem. Soc.*, 2008, **130**, 7546.
57. G. W. Burton, T. Doba, E. Gabe, L. Huges, F. L. Lee, L. Prasad and K. U. Ingold, *J. Am. Chem. Soc.*, 1985, **107**, 7053.
58. A. Wantanbe, N. Noguchi, A. Fujisama, T. Kodama, K. Tamura, O. Cynshi and E. Niki, *J. Am. Chem. Soc.*, 2002, **122**, 5438.
59. L. R. Mahoney and M. A. DaRooge, *J. Am. Chem. Soc.*, 1970, **92**, 4063.
60. J. A. Howard and E. Furimsky, *Can. J. Chem.*, 1973, **51**, 3738.
61. J. H. B. Chenier, E. Furimsky and J. A. Howard, *Can. J. Chem.*, 1974, **52**, 3682.

62. Radical Reaction Rates in Liquids, in *Landolt-Börnstein New Series*, ed. H. Fischer, Springer-Verlag, New York, 1997, vol. 18, subvol. D2.

63. I. W. C. E. Arends, P. Mulder, K. B. Clark and D. D. M. Wayner, *J. Phys. Chem.*, 1995, **99**, 8182.

64. M. Foti, K. U. Ingold and J. Lusztyk, *J. Am. Chem. Soc.*, 1994, **116**, 9440.

65. M. J. Davies, L. G. Forni and R. L. Willson, *Biochem. J.*, 1988, **255**, 513.

66. L. R. Mahoney and M. A. DaRooge, *J. Am. Chem. Soc.*, 1975, **97**, 4722.

67. R. H. Bisby and A. W. Parker, *Arch. Biochem. Biophys.*, 1995, **317**, 170.

68. E. P. L. Hunter, M. F. Desrosiers and M. G. Simic, *Free Radical Bio. Med.*, 1989, **6**, 581.

69. D. W. Snelgrove, J. Lusztyk, J. T. Banks, P. Mulder and K. U. Ingold, *J. Am. Chem. Soc.*, 2001, **123**, 469.

70. P. Mulder, H.-G. Korth, D. A. Pratt, G. A. DiLabio, L. Valgimigli, G. F. Pedulli and K. U. Ingold, *J. Phys. Chem. A*, 2005, **109**, 2647.

71. V. W. Manner, T. F. Markle, J. H. Freudenthal, J. P. Roth and J. M. Mayer, *Chem. Commun.*, 2008, 256.

72. See reference 14. The self exchange rate constants were determined as described in: A. I. Prokof'ev, N. A. Malysheva, N. N. Bubnov, S. P. Solodovnikov, M. I. Kabachnik, *Bull. Acad. Sci. USSR Div. Chem. Sci.*, 1976, **25**, 494.

73. M. Chou, C. Creutz and N. Sutin, *J. Am. Chem. Soc.*, 1977, **99**, 5615.

74. G. Just and Y. Kauko, *Z. Phys. Chem.* 1911, **76**, 601; (b) F. Hein, W. Daniel and H. Schwedler, *Z. Anorg. Chem.*, 1937, **233**, 161; (c) A. H. Webster and J. Halpern, *Trans. Faraday Soc.*, 1957, **53**, 51; (d) J. Halpern, *Adv. Catal.*, 1957, **9**, 302; (d) K. B. Wiberg and G. Foster, *J. Am. Chem. Soc.* 1961, **83**, 423; (e) K. B. Wiberg, in *Oxidation in Organic Chemistry Part A*, ed. K. B. Wiberg, Academic Press, New York, 1965, pp. 69–184.

75. *cf.*, (a) J. T. Groves, *Proc. Nat. Acad. Sci. U. S. A.*, 2003, **100**, 3569; (b) J. T. Groves, *J. Chem. Educ.*, 1985, **62**, 928.

76. (a) *Organic Syntheses by Oxidation with Metal Compounds*, ed. W. J. Mijs and C. R. H. I. de Jonge, Plenum, New York, 1986; (b) D. Arndt, *Manganese Compounds as Oxidizing Agents in Organic Chemistry*, Open Court Publishing, La Salle, IL, 1981.

77. *Cf.*, (a) R. K. Grasselli and J. D. Burrington, *Adv. Catal.*, 1981, **30**, 133; (b) R. K. Grasselli, *J. Chem. Educ.*, 1986, **63**, 216; (c) J. A. Labinger, *Catal. Lett.*, 1988, **1**, 371; (d) G. Centi, F. Trifiro, J. R. Ebner and V. M. Franchetti, *Chem. Rev.*, 1988, **88**, 55.

78. (a) M.-H. Baik, M. Newcomb, R. A. Friesner and S. J. Lippard, *Chem. Rev.*, 2003, **103**, 2385; (b) S. Chakrabarty, R. N. Austin, D. Deng, J. T. Groves and J. D. Lipscomb, *J. Am. Chem. Soc.*, 2007, **129**, 3514; (c) C. Krebs and D. G. Fujimori, C. T. Walsh and J. M. Bollinger Jr., *Acc. Chem. Res.*, 2007, **40**, 484.

79. M. Glickman and J. P. Klinman, *Biochemistry*, 1996, **35**, 12882.

80. N. Nakanishi, F. Takeuchi and M. Tsubaki, *J. Biochem.*, 2007, **142**, 553.

81. D. C. Eisenberg and J. R. Norton, *Isr. J. Chem.*, 1991, **31**, 55.

82. R. M. Bullock, *Comments Inorg. Chem.*, 1991, **12**, 1.

83. (a) V. W. Manner, Concerted Proton-Electron Transfer Reactions of Ruthenium and Cobalt Complexes: Studies on Distance Dependence and Spin Effects, Ph.D Thesis, University of Washington, Seattle, WA, April 2009; (b) V. W. Manner, A. D. Lindsay, E. A. Mader, J. N. Harvey and J. M. Mayer, *Chem. Sci.*, 2011, DOI: 10.1039/c1sc00387a.
84. B. A. Moyer and T. J. Meyer, *J. Am. Chem. Soc.*, 1978, **100**, 3601.
85. A. Haim, *Prog. Inorg. Chem.*, 1983, **30**, 273.
86. L. Eberson, in *Advances in Physical Organic Chemistry*, ed. V. Gold and D. Bethell, Academic, New York, 1982, vol. 18, p. 79.
87. E. A. Mader and J. M. Mayer, *Inorg. Chem.*, 2010, **49**, 3685.
88. J. P. Roth, Intrinsic and Thermodynamic Influences on Hydrogen Atom Transfer Reactions Involving Transition Metal Complexes, Ph.D. Thesis, University of Washington, Seattle, WA, 2000.
89. A. S. Larsen, K. Wang, M. A. Lockwood, G. L. Rice, T.-J. Won, S. Lovell, M. Sadílek, F. Tureek and J. M. Mayer, *J. Am. Chem. Soc.*, 2002, **124**, 10112.
90. For a variety of perspectives on biological hydrogen tunneling see: Quantum catalysis in enzymes: beyond the transition state theory paradigm, a special issue of *Philos. Trans. R. Soc., B*, 2006, **361**, 1293.
91. The xanthene self-exchange rate constant was corrected by a statistical factor of 2 to account for the fact that xanthene has half as many abstractable protons as 9,10-dihydroanthracene. Therefore, $k_{\text{XH/X}}$(xanthene) = $2k_{\text{XH/X}}$(9,10-dihydroanthracene).
92. C. Isborn, D. A. Hrovat, W. T. Borden, J. M. Mayer and B. K. Carpenter, *J. Am. Chem. Soc.*, 2005, **127**, 5794 and references therein.
93. (a) J. M. Tedder, *Angew. Chem., Int. Ed. Engl.*, 1982, **21**, 401. For additional discussions of polar effects, see: (b) W. H. Davis Jr. and W. A. Pryor, *J. Am. Chem. Soc.*, 1977, **99**, 6365; (c) W. A. Pryor, F. Y. Tang, R. H. Tang and D. F. Church, *J. Am. Chem. Soc.*, 1982, **104**, 2885; (d) A. A. Zavitsas and J. A. Pinto, *J. Am. Chem. Soc.*, 1972, **94**, 7390.
94. X. X. Rong, H.-Q. Pan, W. R. Dolblier Jr. and B. E. Smart, *J. Am. Chem. Soc.*, 1994, **116**, 4521.
95. M. K. Ludlow, A. V. Soudackov and S. Hammes-Schiffer, *J. Am. Chem. Soc.*, 2009, **131**, 7094.
96. E. Hatcher, A. V. Soudackov and S. Hammes-Schiffer, *J. Am. Chem. Soc.*, 2004, **126**, 5663.

CHAPTER 2

A Transition-State Perspective of Proton-Coupled Electron Transfers

LUIS G. ARNAUT

Department of Chemistry, University of Coimbra, 3004-535 Coimbra, Portugal

2.1 Introduction

Hydrogen-atom transfer (HAT) reactions:

$$AH + B \rightarrow A + HB \tag{I}$$

are among the best known chemical reactions, with extensive kinetic databases covering a wide variety of systems and potential-energy surfaces with chemical accuracy available for prototypical systems. The chemical understanding of the molecular factors that control the reactivity of these systems is also particularly advanced.[1–4] On the other hand, the apparently similar transfer of a proton from an acid to a base coupled with the transfer of an electron between other donor–acceptor molecular groups or entities:

$$\text{(II)}$$

RSC Catalysis Series No. 8
Proton-Coupled Electron Transfer: A Carrefour of Chemical Reactivity Traditions
Edited by Sebastião Formosinho and Mónica Barroso
© Royal Society of Chemistry 2012
Published by the Royal Society of Chemistry, www.rsc.org

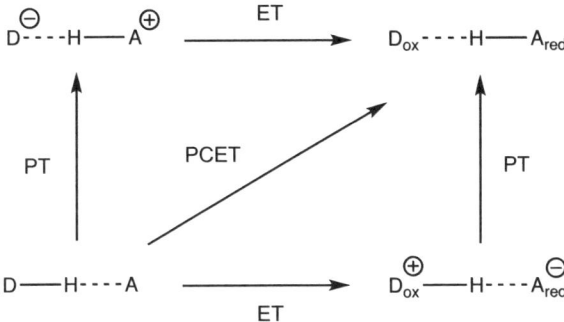

Figure 2.1 Square diagram for sequential PT–ET or ET–PT and for synchronous PCET reactions.

remains a subject of intense scrutiny and some controversy. The distinct feature of coupled proton-electron transfers (CPET) is that the overall transfer of a hydrogen atom occurs in an elementary step in which the electron and the proton are transferred from different orbital sites on a donor to different orbital sites on an acceptor. The square scheme in Figure 2.1 illustrates the difference between concerted and sequential transfers of proton and electron between two hydrogen-bonded species. On the other hand, both in HAT and in CPET the transfer of the proton and of the electron are synchronous, and both are currently designated proton-coupled electron transfers (PCET).

The renewed interest in PCET recognizes its relevance for the understanding of solar energy conversion of natural and artificial systems,[5–7] enzyme catalysis,[8] electrochemistry[9] and chemical reactivity.[10] The challenge posed by CPET is that it combines in an elementary step the intrinsically adiabatic proton transfer with an intrinsically non-adiabatic electron transfer. The PT reaction path connects strongly interacting states and is an adiabatic reaction. On the contrary, ET between spatially separated orbitals or orbitals with different symmetries, occurs with weak coupling between the initial and final electronic states and is nonadiabatic. This is illustrated in Figure 2.2, where the surfaces of the initial and final states of the electron transfer event are allowed to cross with weak interaction.

A concerted reaction path requires changes of molecular geometries in various regions of the reactants at the same time. This gives more contributions to the reaction barrier than sequential steps that normally have an elementary step with a higher barrier that determines the kinetics of the reaction mechanism. However, the concerted path may have a lower energy saddle point if it avoids high-energy intermediates. This is the case illustrated in Figure 2.2.

The energy contributions to the barrier of the PCET coming from independent changes in geometry that promote ET or PT can be expected to be additive. A physically meaningful model to describe PCET should be able to describe, with the same framework, the barriers of sequential PT and ET reactions in Mechanism (II), and also the barriers of HAT following

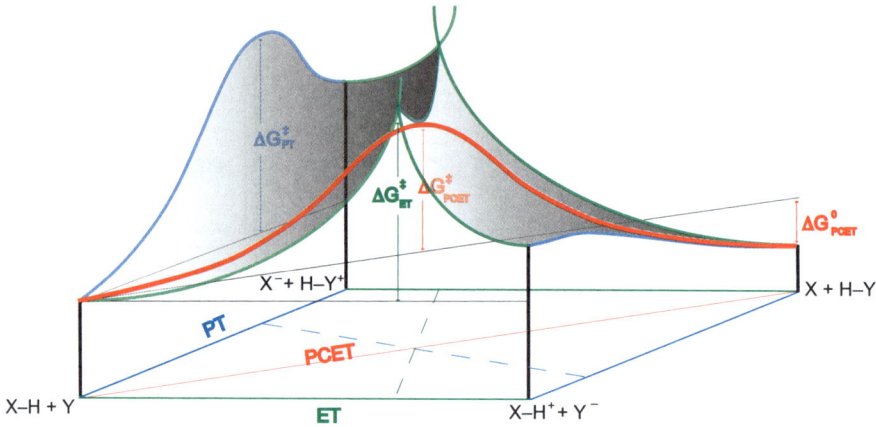

Figure 2.2   Surfaces involved in sequential and concerted proton and electron transfers, illustrating the reaction coordinate for CPET (red), the diabatic crossing of the surfaces for the ET reaction (green) and the adiabatic surfaces for PT (blue). The rate-determining step of each sequential mechanism is given by the largest barrier of the two sequential reactions ($\Delta G_{\mathrm{PT}}^{\ddagger}$ for the PT $\rightarrow$ ET mechanism and $\Delta G_{\mathrm{ET}}^{\ddagger}$ for the ET $\rightarrow$ PT mechanism). The dashed lines connect the locations of the transition states in the PT coordinates (blue) or in the ET coordinates (green). The PCET avoids high-energy intermediates and, consequently, high-energy barriers. The saddle point of the CPET reaction path is shown to coincide with the intersection seam of the diabatic surfaces.

Mechanism (I). In this chapter we merge the approaches of the Interacting/Intersecting State Model (ISM) to atom and proton transfer reactions[4] with treatment of electron transfer reactions by this same model.[11] Transition-state energies and effective reaction frequencies are calculated and used to obtain the rate constants of representative CPET.

2.2   Theory

2.2.1   Hydrogen Atom Transfers

According to the transition-state theory, the rate constant of a prototypical HAT

$$A - H + B \rightleftarrows \{A \cdots H \cdots B\}^{\ddagger} \xrightarrow{\nu^{\ddagger}} A + H - B \qquad (\mathrm{III})$$

takes the form, for the particular case where A and B are monoatomic species:[3]

$$k_{\mathrm{AHB}} = \kappa_{\mathrm{tun}} \sigma \frac{k_{\mathrm{B}} T}{h} \frac{Q^{\ddagger}}{Q_{\mathrm{A}} Q_{\mathrm{HB}}} \exp\left(-\frac{\Delta V_{\mathrm{ad}}^{\ddagger}}{RT}\right) \qquad (2.1)$$

where κ_{tun} is a tunnelling correction, σ a statistical factor, Q_i represents the partition functions of the triatomic transition-state or of the reactants, and $\Delta V^{\ddagger}_{ad}$ is the vibrationally-adiabatic barrier. The partition functions of poly-atomic reactants contain many terms that cancel with analogous terms in the transition-state. We have shown that the ratio of partition functions in HAT following Mechanism (III) can be estimated from the values of the triatomic transition-state $\{A \cdots H \cdots B\}^{\ddagger}$, diatomic reactants A–H and monoatomic B, multiplied by $(q_v/q_r)^r$ taking the value of the rotational partition function as 3 times that of the vibrational one[12] and making $r = 0$ for atom + polyatomic, $r = 2$ for diatomic + polyatomic and $r = 3$ polyatomic + polyatomic systems.

$$k_{TST} = \kappa_{tun}(T)\sigma \frac{k_B T}{h} \left(\frac{1}{3}\right)^r \frac{Q^{\ddagger}}{Q_A Q_{HB}} \exp\left(-\frac{\Delta V_{ad}^{\ddagger}}{RT}\right) \tag{2.2}$$

Pre-exponential factors for atom + polyatomic systems are in the order of magnitude of 10^{10} mol^{-1} dm^3 s^{-1}, and decrease to 5×10^8 mol^{-1} dm^3 s^{-1} for polyatomic + polyatomic systems. The semi-classical tunnelling correction κ_{tun} strongly depends on the temperature.[13] For the simplest and most accurately known HAT, $H + H_2$, κ_{tun} is in the range 7–11 at 300 K.[12]

The vibrationally-adiabatic barrier can be obtained adding the zero-point energy to the classical energy at each point along the reaction path, taken as the bond order of the product HB under the assumption that the bond order is conserved along the reaction coordinate,[14] $n = n_{HB} = 1 - n_{HA}$,

$$V_{ad}(n) = V_{cl}(n) + \sum_i \left(\frac{1}{2}hc\bar{\nu}_i\right) \tag{2.3}$$

where $\bar{\nu}_i$ are the vibration frequencies of the normal modes orthogonal to the reaction coordinate. The frequencies of the linear triatomic transition-state have been estimated from Wilson's equation while neglecting of the interaction between bending and stretching, using fractional bonds in the $\{A \cdots H \cdots B\}^{\ddagger}$ transition-state and a switching function to provide the correct asymptotic limits. The form selected for the switching function is[12]

$$\begin{aligned} y(n) &= \cosh\left[\frac{l_{HA}}{l_{HB} + l_{HA}} \frac{\ln(n^{\ddagger})}{\ln(2n)}\right]^{-1} & \text{for } n < 0.5 \\ y(n) &= 0 & \text{for } n = 0.5 \\ y(n) &= \cosh\left[\frac{l_{HA}}{l_{HB} + l_{HA}} \frac{\ln(1 - n^{\ddagger})}{\ln(2(1 - n))}\right]^{-1} & \text{for } n > 0.5 \end{aligned} \tag{2.4}$$

An empirical linear relation between symmetric stretching and bending frequencies in triatomic systems was employed to estimate the bending frequency from the symmetric stretching frequency.

The classical reaction path of ISM is a linear interpolation between the Morse curves of HA and HB along the reaction coordinate using the bond order of the products

$$V_{cl}(n) = (1 - n)V_{HA} + nV_{HB} + n\Delta V^0 \tag{2.5}$$

where ΔV^0 is the classical reaction energy and the Morse curves have the form

$$
\begin{aligned}
V_{HA} = D_{e,HA}\left\{1 - \exp\left[-\beta_{HA}\left(l_{HA}^{\ddagger} - l_{HA,eq}\right)\Big/m\right]\right\}^2 \\
V_{HB} = D_{e,HA}\left\{1 - \exp\left[-\beta_{HB}\left(l_{HB}^{\ddagger} - l_{HB,eq}\right)\Big/m\right]\right\}^2
\end{aligned}
\tag{2.6}
$$

and the electrophilicity index m is the ratio between the negative of the electronic chemical potential, μ_{el}, and the chemical hardness, η_{el},[15]

$$m = \frac{-\mu_{el}}{\eta_{el}} = \frac{I_P + E_A}{I_P - E_A} \tag{2.7}$$

where I_P is the ionization potential and E_A is the electron affinity of A or B. The value of m can be calculated from the electronic parameters of A or B, but each transition-state can only have one value of m. For proton transfers we assume that the value of m at the transition-state is determined by the electronic properties of the species with the lowest I_P, because it is easier to delocalize the electrons of that species. For hydrocarbons, which have low electron affinities, we simply take $m = 1$ along the entire reaction coordinate. Otherwise, a switching function similar to eqn (2.4) also provides the asymptotic limits of $m = 1$ for the reactants and products and the value given by eqn (2.7) at the transition state.[16]

The bond extensions from equilibrium to transition-state configurations are given by a generalization of the Pauling relation between equilibrium bond lengths and bond orders,[17]

$$
\begin{aligned}
l_{HA}^{\ddagger} - l_{HA,eq} = -a'_{sc}\left(l_{HA,eq} + l_{HB,eq}\right)\ln\left(n_{HA}^{\ddagger}\right) \\
l_{HB}^{\ddagger} - l_{HB,eq} = -a'_{sc}\left(l_{HA,eq} + l_{HB,eq}\right)\ln\left(n_{HB}^{\ddagger}\right)
\end{aligned}
\tag{2.8}
$$

The scaling by $a'_{sc}(l_{AB,eq} + l_{BC,eq})$ reflects the fact that longer bonds will stretch out more from equilibrium to the transition-state configurations than shorter ones, and that two bonds are implicated in the transition-state. The value of a'_{sc} was obtained from the bond extension of the $H + H_2$ system from the equilibrium to the transition-state bond lengths,[18] $a'_{sc} = 0.182$.

2.2.2 Proton Transfers in Hydrogen-Bonded Systems

CPET usually takes place in H-bonded systems. In such systems we have to take into consideration the presence of the H-bond

$$A - H + B \underset{k_{-a}}{\overset{k_a}{\rightleftharpoons}} (A - H \cdots B) \xrightarrow{\nu_{(AB)}} \{A \cdots H \cdots B\}^{\ddagger}$$

$$\xrightarrow{\nu_{(AB)}} (A^- \cdots H-B^+) \rightleftharpoons A^- + H - B^+ \qquad (IV)$$

The formation of the H-bond can be regarded as an advancement along the reaction coordinate,[16] making $n_{HA} < 1$ and $n_{HB} > 0$. The bond orders of the H-bonded reactant and product complexes can be related to the corresponding bond lengths in the same form as eqn (2.8),

$$n_{HA}{}^* = \exp\left[-\frac{l_{HA}{}^* - l_{HA,eq}{}^*}{a'\left(l_{HA}{}^* - l_{HA,eq}{}^*\right)}\right] \qquad (2.9a)$$

$$n_{HB}{}^* = \exp\left[-\frac{l_{HB}{}^* - l_{HB,eq}{}^*}{a'\left(l_{HB}{}^* - l_{HB,eq}{}^*\right)}\right] \qquad (2.9b)$$

and the bond lengths of the AH and HB bonds in the H-bonded complexes ($l_{HB,eq}{}^*$ and $l_{HB,eq}{}^*$) can be related to the strength of the H-bond using the Lippincott–Schroeder (LS) potential.[19]

The LS potential relates the H-bond binding energies (D_{AB}) to the AB equilibrium distances ($l_{AB,eq}$) and to AB stretching frequency ($\bar{\nu}_{AB}$) in the $B \cdots H-A$ hydrogen-bonded complex. It is the sum of four terms[19]

$$V_{LS} = V_{cov,AH} + V_{cov,HB} + V_{rep} + V_{el} \qquad (2.10)$$

The first two terms represent the covalent interactions present in the H-bond and have the forms

$$V_{cov,HA} = D_{HA}\left\{1 - \exp\left[-\frac{\zeta\left(l_{HA} - l_{HA,eq}\right)^2}{2l_{HA}}\right]\right\} \qquad (2.11a)$$

$$V_{cov,HB} = D_{HB^+}\left\{1 - \exp\left[-\frac{\zeta\left(l_{HB^+} - l_{HB^+,eq}\right)^2}{2l_{HB^+}}\right]\right\} - D_{HB^+} \qquad (2.11b)$$

where

$$\begin{aligned}\zeta_{HA} &= f_{HA}l_{HA,eq}/D_{HA} \\ \zeta_{HB^+} &= f_{HB^+}l_{HB^+,eq}/D_{HB^+}\end{aligned} \qquad (2.12)$$

The H–A bond is essentially a slightly stretched HA bond and the corresponding harmonic force constant (f_{HA}), equilibrium bond length ($l_{HA,eq}$) and binding energy (D_{HA}) can be taken from the isolated molecule. However, for the $B \cdots H$ bond the quantities f_{HB^+}, $l_{HB^+,eq}$ and D_{HB^+} are not readily available. For this bond, Lippincott and Schroeder introduced the approximation

$$\zeta_{HB^+} = g\zeta_{HA} \tag{2.13}$$

and the assumption

$$D_{HB^+} = D_{HA}/g \tag{2.14}$$

which implies f_{HB+} and $l_{HB+,eq}$ can be taken from the unperturbed HB bond ($f_{HB+} \approx f_{HB}$ and $l_{HB+,eq} \approx l_{HB}$), and showed that the parameter g is transferable to all A–H$\cdots$B systems, $g = 1.45$. This provides all the relations necessary to calculate the covalent contributions from the data on the unperturbed HA and HB bonds.

The repulsive term V_{rep} was expressed as a negative exponential and the electrostatic term V_{el} as a negative power of the AB distance, l_{AB}. Both these terms involve empirical constants and were modified to reduce the number of such constants,

$$V_{rep} + V_{el} = A\left[\exp(-bl_{AB}) + \frac{l_{AB}^*}{2l_{AB}}\exp(-bl_{AB}^*)\right] \tag{2.15}$$

where $b = 4.8$ Å.[19] The constant A is given by the first derivative of the LS potential

$$A = \frac{D_{HB^+}\zeta_{HB^+}}{2}\left[1 - \left(\frac{l_{HB^+}}{l_{AB}^* - l_{AB}}\right)^2\right]\frac{\exp\left(-\zeta_{HB^+}\left(l_{HB^+} - l_{HB^+,eq}\right)^2/2l_{HB^+}\right)}{\exp(-bl_{AB}^*)}$$
$$\times \left(b - \frac{1}{2l_{AB}^*}\right) \tag{2.16}$$

and the force constant of the AB bond can be obtained fom the second derivative

$$f_{AB} = B + A\exp(-bl_{AB}^*)\frac{(bl_{AB}^*)^2 - 1}{(l_{AB}^*)^2} \tag{2.17a}$$

$$B = \frac{D_{HB^+}\zeta_{HB^+}}{l_{HB^+}}\exp\left(-\frac{\zeta_{HB^+}\left(l_{HB^+} - l_{HB^+,eq}\right)^2}{2l_{HB^+}}\right)$$
$$\times\left[\left(l_{HB^+,eq}\right)^2 - \frac{\zeta_{HB^+}\left(l_{HB^+} - l_{HB^+,eq}\right)^4}{4l_{HB^+}}\right] \tag{2.17b}$$

The relevance of the force constant lies in its relation with the AB vibrational frequency

$$\nu_{e(AB)} = \frac{1}{2\pi}\sqrt{\frac{f_{AB}}{\mu_{AB}}} \tag{2.18}$$

This frequency is a measure of the PT reaction frequency in H-bonded systems. The frequency factor of the PT step is the product between the stationary concentration of the H-bonded collision complexes for unit concentration of the two reactants, times the frequency of the reactive hydrogen bond. This is identical to the product of the association constant K_c ($= k_a/k_{-a}$) by the frequency of the promoting mode of PT in H-bonded systems, which is the stretching frequency of the heavy atoms ($\nu_{e(AB)}$),[20]

$$k_{PT} = \kappa(T)c\bar{\nu}_{e(AB)}K_c \exp\left(-\frac{\Delta V_{ad}^{\ddagger}}{RT}\right) \qquad (2.19)$$

Typical values of $\bar{\nu}_{e(AB)}$ range from 100 to 1000 cm^{-1}, hence the reaction frequency should be between 3×10^{12} and 3×10^{13} s^{-1}, which brackets the classical reaction frequency of transition-state theory, 6×10^{12} s^{-1} at room temperature. The difference in the pre-exponential factors of eqn (2.2) and (2.19) comes from the difference between the partition functions *versus* the value of K_c. K_c accounts for the fraction of H-bonded systems, which are the only reactive systems. Earlier estimates of k_a/k_{-a} range between 0.1 and 0.4 M^{-1} in water for carboxylic acids and their conjugated bases,[21] but can be expected to depend on the nature of the proton donor and acceptor and of the solvent. The assumption that only H-bonded systems react has been extensively discussed by Ingold and related to the strength of the H-bond in different solvents.[22] Solvent kinetic effects can be included in the rate calculations using the H-bond ability of the acid and the H-bond acceptor properties of the solvent.[23]

In addition to the changes in the pre-exponential factor, the formation of an H-bond typically reduces the height and width of the reaction barrier. Narrow barriers are very permeable to tunnelling by light species such as the hydrogen atom or the proton. When nuclear tunnelling becomes the dominant process, it is possible to formulate the HAT rate constant as a non-adiabatic process and express it in terms of the Gamov formula for one-dimensional nuclear tunnelling. Seminal work by Formosinho[24] showed that photochemical HAT can be regarded as radiationless transitions with tunnelling rates given by

$$k_{tunnel} = \nu \exp\left[-\frac{\gamma}{\hbar}d\sqrt{\mu\Delta E^{\ddagger}}\right] \qquad (2.20)$$

where ν is the effective frequency with which the particle of mass μ hits the barrier, $\Delta E^{\ddagger}$ is the barrier height and d its width. The value of γ is $2^{3/2}$ for a rectangular barrier, or $\gamma = \pi/2^{1/2}$ for a parabolic barrier or $\gamma = 2^{1/2}$ for a barrier formed by the intersection of two parabola. Jortner and Ulstrup later showed that the nuclear Franck–Condon factors of atom transfer reactions can be expressed in a form identical to the equation above with $\gamma = 2^{1/2}$ when $k_BT \gg h\nu$,[25] and a large number of photochemical HATs have been calculated as radiationless transitions.[26] Hydrogen bonding to the solvent may convert these HATs into electron transfer from B to electronically-excited A coupled with proton transfer to the solvent,[27] which is a form of CPET.

The treatment of atom-transfer reactions as radiationless transitions between classical potential energy surfaces was further developed by Ulstrup,[28] who expressed the single-mode rate as

$$w = \kappa_{ad} \frac{\omega_{eff}}{2\pi} \exp\left(-\frac{E_{\ddagger}}{k_B T}\right) \tag{2.21}$$

where ω_{eff} is the effective frequency of the classical modes, and the transmission coefficient κ_{ad} and the energy barrier $E_{\ddagger}$ take different values according to the degree of non-adiabaticity of the reaction. In the limit of an adiabatic reaction, $\kappa_{ad} = 1$ and $E_{\ddagger}$ is the classical energy barrier (*i.e.*, $\Delta V^{\ddagger}_{ad}$ minus zero-point energy corrections) and we recover eqn (2.19). In the limit of totally non-adiabatic processes, the transmission coefficient takes the value of the classical probability of transition given by the Landau–Zener theory,

$$\kappa_{ad} = 1 - \exp\left[-\frac{2\pi(\Delta\varepsilon_{ad}/2)^2}{\hbar v_c |s_1 - s_2|}\right] \approx \frac{2\pi(\Delta\varepsilon_{ad}/2)^2}{\hbar v_c |s_1 - s_2|} \tag{2.22}$$

where v_c is the velocity of passage over the crossing point, s_i are the slopes of the potential energy curves at that point, and $\Delta\varepsilon_{ad}$ is the splitting of the zero-order electronic potential surfaces with respect to the coordinates of the slow nuclear subsystem at the crossing point. The value of $E_{\ddagger}$ in the non-adiabatic limit is the activation energy determined by the modes of the slow subsystem.[28]

2.2.3 Electron Transfers

The first influential understanding of the factors that control the rate of electron transfer from a donor (D, or reduced species) to an acceptor (A, or oxidized species) in a solvent of refractive index n_D and dielectric constant ε,

$$D + A \xrightarrow{k_{ET}} D^+ + A^- \tag{V}$$

was presented by Marcus in the 1960's in terms of solvent and molecular reorganization energies.[29–31] The solvent reorganization energy was obtained using a dielectric continuum model

$$\lambda_s = e^2 \left(\frac{1}{2r_D} + \frac{1}{2r_A} - \frac{1}{r}\right)\left(\frac{1}{n_D^2} - \frac{1}{\varepsilon}\right) \tag{2.23}$$

where r_D and r_A are the radii of spheres representing the donor and acceptor, respectively, $r = r_D + r_A$ is the D–A centre-to-centre separation distance, and a strong dependence of λ_s on the polarity of the solvent is expected. The reorganization energy associated with each intramolecular configuration change was expressed using a modification of the George–Griffith model[32]

$$\lambda_v = 4\sum_i \frac{1}{2}\left[f_{ox(i)}\left(l_{(i)}^* - l_{ox(i)}\right)^2 + f_{red(i)}\left(l_{(i)}^* - l_{red(i)}\right)^2\right] \tag{2.24}$$

for the case of a critical configuration for electron transfer, for the ith bond, of

$$l_{(i)}^* = \frac{f_{ox(i)}l_{ox(i)} + f_{red(i)}l_{red(i)}}{f_{ox(i)} + f_{red(i)}} \tag{2.25}$$

which leads to the structural reorganization energy

$$\lambda_v = \sum_i \frac{1}{2} \frac{f_{ox(i)}f_{red(i)}}{f_{ox(i)} + f_{red(i)}} \left(l_{ox(i)} - l_{red(i)}\right)^2 \tag{2.26}$$

The reorganization energy of the molecular bonds is critically dependent on the difference between the equilibrium bond lengths of each bond in its oxidized (l_{ox}) or reduced (l_{red}) states. The additivity of reorganization energies is also extended to the solvent, and the total reorganization energy is $\lambda = \lambda_s + \lambda_v$. Later, Marcus expressed cross-reaction rates in terms of the self-exchange rates and obtained the quadratic energy dependence of the rates on the reaction free energy[33]

$$\Delta G^* = \frac{\lambda}{4} \left(1 + \frac{\Delta G^0}{\lambda}\right)^2 \tag{2.27}$$

where the reorganization energy λ is assumed constant along the reaction series and taken as $4\Delta G_0^{\ddagger}$, *i.e.*, four times the barrier at $\Delta G^0 = 0$.

A classical expression for the ET rate constant can be derived expressing the rate constant for the transition from the initial to the final state as the product between the probability of conversion at the crossing point and the frequency of passage over the crossing point

$$k_{cl} = 2\nu P_{12} \tag{2.28}$$

where the factor of 2 accounts for the fact that for a nonadiabatic process the intersection region is passed twice during a period of a vibrational motion, and the frequency of passage is related to the vibrational frequency of the promoting modes of the transition, $\nu = \omega_v/(2\pi)$. The classical probability of transition is given by the Landau–Zener theory, which for small conversion probabilities can also be written, following eqn (2.22), as

$$p_{LZ} = \frac{2\pi|V_{12}|^2}{\hbar v_c |s_1 - s_2|} \tag{2.29}$$

For electronic states represented by unidimensional harmonic oscillators with the same vibrational frequency, this leads to

$$p_{LZ} = \frac{2\pi}{\hbar} \frac{|V_{12}|^2}{\omega} \frac{1}{\sqrt{4\lambda}} \left(E - E^{\ddagger}\right)^{-\frac{1}{2}} \tag{2.30}$$

which, after averaging over a Boltzmann distribution of energies and using the quadratic dependence of the barrier on the reaction energy, gives the classical rate[3]

$$k_{cl} = \frac{2\pi}{\hbar}|V_{12}|^2 \frac{1}{\sqrt{4\pi\lambda k_B T}} \exp\left[-\frac{(\Delta E^0 - \lambda)^2}{4\lambda k_B T}\right] \tag{2.31}$$

The reaction energy ΔE^0 is defined here as the difference in electronic energy between the initial (donor) and final (acceptor) states in their ground vibrational states, and is positive when the energy of the donor is higher than that of the acceptor, *i.e.*, $\Delta E^0 > 0$ for exergonic processes. It is more common to write the classical ET rate constant in terms of free-energies

$$k_{cl} = \frac{2\pi}{\hbar}|V_{12}|^2 \frac{1}{\sqrt{4\pi\lambda k_B T}} \exp\left[-\frac{(\Delta G^0 + \lambda)^2}{4\lambda k_B T}\right]$$

$$= \frac{2\pi}{\hbar}|V_{12}|^2 \frac{1}{\sqrt{4\pi\lambda k_B T}} \exp\left[-\frac{\Delta G^*}{k_B T}\right] \tag{2.32}$$

The dependences of the ET rate constant on the solvent polarity and on the reaction free-energy predicted by Marcus theory have been thoroughly tested. The solvent dependence of self-exchange rates ($\Delta G^0 = 0$) revealed that the dielectric continuum model overestimates λ_s in polar solvents.[11,34–36] The free-energy dependence of ET rates should lead to an inverted parabola when $\ln(k_{ET})$ is plotted as a function of ΔG^0, with a maximum at $\Delta G^0 = -\lambda$. This was observed in rigid donor–acceptor systems and in ion-pairs, but even in such cases the "inverted region" of the very exothermic reactions has a less pronounced free-energy dependence than the "normal region" of the less exothermic reactions.[35,37–44]

The asymmetric dependence of the activation energy on the driving force (energy-gap law) of ET reactions can be reproduced using the golden rule of time-dependent perturbation theory. The golden rule allows for the calculation of the transition rate from an initial to a final electronic state subject to a weak perturbation applied for a short period of time. The ET rate constant was first expressed in terms of the golden rule in the 1970's[45–47]

$$k_{gr} = \frac{2\pi}{\hbar}|V_{12}|^2(FCF) \tag{2.33}$$

where the Franck–Condon factor (*FCF*) is associated with the molecular structures of electron donor and acceptor, and the electronic coupling (V_{12}) is associated with the overlap of the electronic wavefunctions of donor and acceptor. The Franck–Condon factor contains the dependence of the reaction rate on the reaction energy, and the electronic coupling its dependence on the donor–acceptor distance, $V^2 = V_0^2\exp(-\beta r_{DA})$, where r_{DA} is the donor–acceptor contact distance and β the distance decay factor. *FCF* is the vibrational overlap integral between initial and final states. Within the displaced harmonic oscillator approximation, and when the initial (donor, D) and final (acceptor, A) states are each represented by one single high frequency, this overlap is given by

$$FCF = \frac{1}{\sqrt{2\pi(\sigma_D^2 + \sigma_A^2)}} \sum_{j=0}^{\infty} \sum_{k=0}^{\infty} e^{-S_D} e^{-S_A} \frac{S_D^k}{k!} \frac{S_A^j}{j!} \exp\left[-\frac{[\Delta E_{DA} + k\hbar\omega_D + j\hbar\omega_A]^2}{2(\sigma_D^2 + \sigma_A^2)}\right]$$

(2.34)

where σ^2 is the width of the Gaussian spectral functions,

$$\Delta E_{DA} = \Delta E^0 - (\lambda_D + \lambda_A)$$

(2.35)

where λ_D and λ_A are the reorganization energies of donor and acceptor, and S is the electron-vibration coupling strength for the distinct vibration mode ω_v,

$$S = \frac{\lambda_v}{\hbar\omega_v} = \frac{\frac{1}{2}f(l_{ox} - l_{red})^2}{\hbar\omega_v} = \left(\frac{l_{ox} - l_{red}}{2\sigma_v}\right)^2$$

(2.36)

which is also the change of the equilibrium configuration of the vibrational mode expressed in units of the root-mean-square displacement at the zero-point energy.[48]

It has been shown that even when ω_D and ω_A differ by as much as 10% the double sum can be replaced by a single sum[49]

$$FCF = \frac{1}{\sqrt{4\pi\sigma^2}} \sum_{n=0}^{\infty} e^{-S} \frac{S^n}{n!} \exp\left[-\frac{(\Delta E^0 - \lambda + n\hbar\omega_v)^2}{4\sigma^2}\right]$$

(2.37)

where $S = S_A + S_D$, $\lambda = \lambda_A + \lambda_D$ $\omega_v = \omega_A = \omega_D$ and $\sigma = \sigma_A = \sigma_D$. Replacing the equation above in the expression for the golden rule, leads to the formula derived by Jortner[50] for the rate constant of ET under the approximation that $\hbar\omega_s << k_B T << \hbar\omega_v$, where $\hbar\omega_s$ refers to a lattice frequency (< 100 cm^{-1}), the reorganization energy λ is also associated with such low frequencies, as well as the widths 2σ of the spectral functions. It became common to employ a semiclassical approximation and write $\sigma^2 = \lambda k_B T$,[51,52] where now λ is the total reorganization energy, and express the ET rate constant as

$$k_{gr} = \frac{2\pi}{\hbar} |V_{12}|^2 \frac{1}{\sqrt{4\pi\lambda k_B T}} \sum_{n=0}^{\infty} e^{-S} \frac{S^n}{n!} \exp\left[-\frac{(\Delta E^0 - \lambda + n\hbar\omega_v)^2}{4\lambda k_B T}\right]$$

(2.38)

The asymmetry of the inverted region increases when the value of S increases, *i.e.*, when the displacement between initial and final states increases. Both the classical and the quantum mechanical models for ET described above assume that the electronic states cross at a given energy, which is the energy of the activated state ΔG^* (or $\Delta E^{\ddagger}$, neglecting entropy changes). However, the transfer of an electron according to Mechanism (V) involves two reactants and two products, hence 4 electronic states. The crossing point is no longer rigorously defined and is replaced by the reorganization energy λ, which is not calculated from the energy at the crossing point. This reorganization energy is calculated from the displacement of the electronic states in each species.

Alternatively, the classical ET rate constant can be formulated as[11]

$$k_{ISM} = \nu_{eff} K_c \exp\left(-\frac{\Delta G_{et}^{\ddagger}}{RT}\right) \tag{2.39}$$

where ν_{eff} is the effective frequency for ET and a good estimate of the energy of the crossing point can be obtained using a high frequency ω_v representative of the dominant promoting modes and the sum of the bond length changes from the equilibrium to the transition-state configuration, shown in Figure 2.3. According to ISM, this sum of bond length changes for an isothermic reaction is given by

$$d = \sum_i \left(\left|l_i^{\ddagger} - l_{ox(i)}\right| + \left|l_i^{\ddagger} - l_{red(i)}\right|\right) = \frac{a_{cl}' \ln(2)}{n^{\ddagger}}(l_{ox} + l_{red}) \tag{2.40}$$

where $a_{cl}' = 0.156$ is the classical constant analogous to the semiclassical a_{sc}' discussed above, and the transition-state bond order $n^{\ddagger}$ for an electron transfer is just the average of the bond orders of the reactants because the bond orders do not

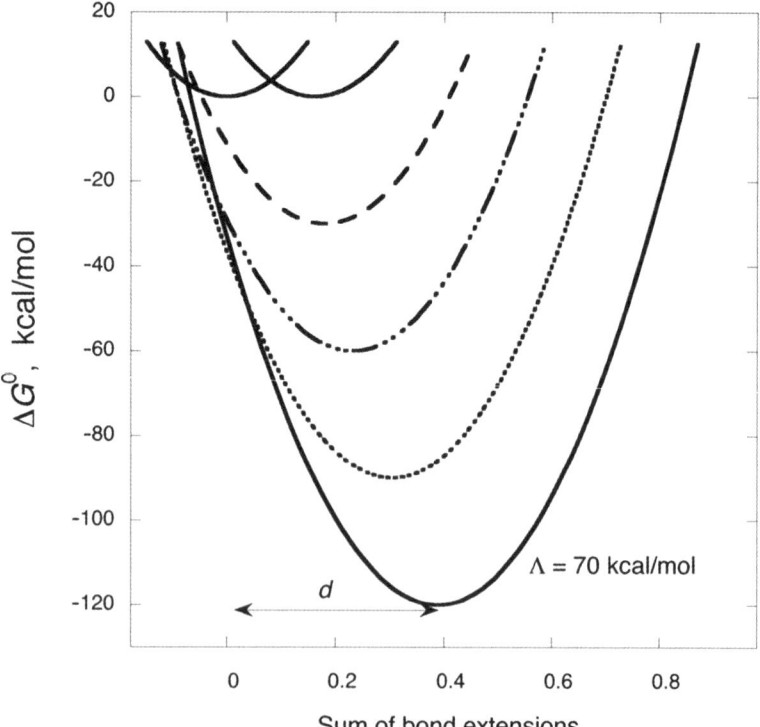

Figure 2.3 Reaction coordinate of ISM for electron transfers.

change along the reaction coordinate of an ET. The corresponding expression for $\Delta G^0 \neq 0$ is

$$d = \frac{a'_{cl}}{2n^{\ddagger}} \ln \left\{ \frac{1 + \exp\left(\sqrt{2n^{\ddagger}} \Delta G^0 / \Lambda\right)}{1 - \left[1 + \exp\left(\sqrt{2n^{\ddagger}} \Delta G^0 / \Lambda\right)\right]^{-1}} \right\} (l_{ox} + l_{red}) \qquad (2.41)$$

where Λ is an empirical parameter that makes the displacement increase with the reaction exothermicity. The empiricism of the dynamic parameter Λ masks the oversimplification of the ISM reaction coordinate to one single high-frequency mode representing the reactants and another one representing the products. The increase in the reaction exothermicity may increase the displacement of the selected high-frequency mode as well as increase the population of other vibrational modes. The correct description of the reaction coordinate of more exothermic reactions requires more vibrational modes and larger reorganization energies. The value of Λ controls the asymmetry of the inverted region, just like the value of S in the golden rule expression. Its precise value is only important for very exothermic ETs.

The value of the effective frequency ν_{eff} depends on the degree of adiabaticity of the reaction. The electronic frequency (*i.e.*, the frequency of the movement of an electron in a molecular orbital, $\nu_{el} = 10^{15}$ s^{-1}) is much higher than the nuclear frequency ($\nu_v = c\bar{\nu}_v$ is between 6×10^{12} and 5×10^{13} s^{-1} when $\bar{\nu}_v$ ranges from 200 to 1500 cm^{-1}) and the electrons have enough time to adjust to the positions of the nuclei. In this case ($\nu_{el} \gg \nu_v$), the frequency of passage over the critical configuration for ET is dictated by the vibrational frequency of the promoting modes, $\nu_{eff} \approx \nu_v$. However, when the probability of ET is impaired by the physical separation between electron donor and acceptor orbital, by spin changes or by orbital symmetry, the ET step becomes strongly nonadiabatic and the effective frequency is controlled by the value of the electronic nonadiabatic factor ($\chi_{el} \ll 1$ for a nonadiabatic ET). We have discussed the shift from adiabatic to non-adiabatic electron transfers in terms of the effective reaction frequencies and showed that the vectors of the nuclear and electronic frequencies in the plane defined by the PT and ET coordinates of Figure 2.2 combine to give[11]

$$\chi_{eff} = \frac{\chi_{el} \nu_{el}}{\nu_v} \sin\left[\arctan\left(\frac{\nu_v}{\chi_{el}\nu_{el}}\right)^2\right] + \cos\left[\arctan\left(\frac{\nu_v}{\chi_{el}\nu_{el}}\right)^2\right] \qquad (2.42)$$

The dependence of χ_{eff} on $\nu_v/(\chi_{el}\chi_{el})$ is presented in Figure 2.4. The ET rate constant can be explicitly written in terms of this electronic non-adiabatic factor

$$k_{ISM} = \chi_{eff} \nu_v K_c \exp\left(-\frac{\Delta G^{\ddagger}}{RT}\right) \qquad (2.43)$$

The degree of nonadiabaticity of ET, and hence the value of χ_{el}, is critically dependent on the distance between electron donor and acceptor and on the

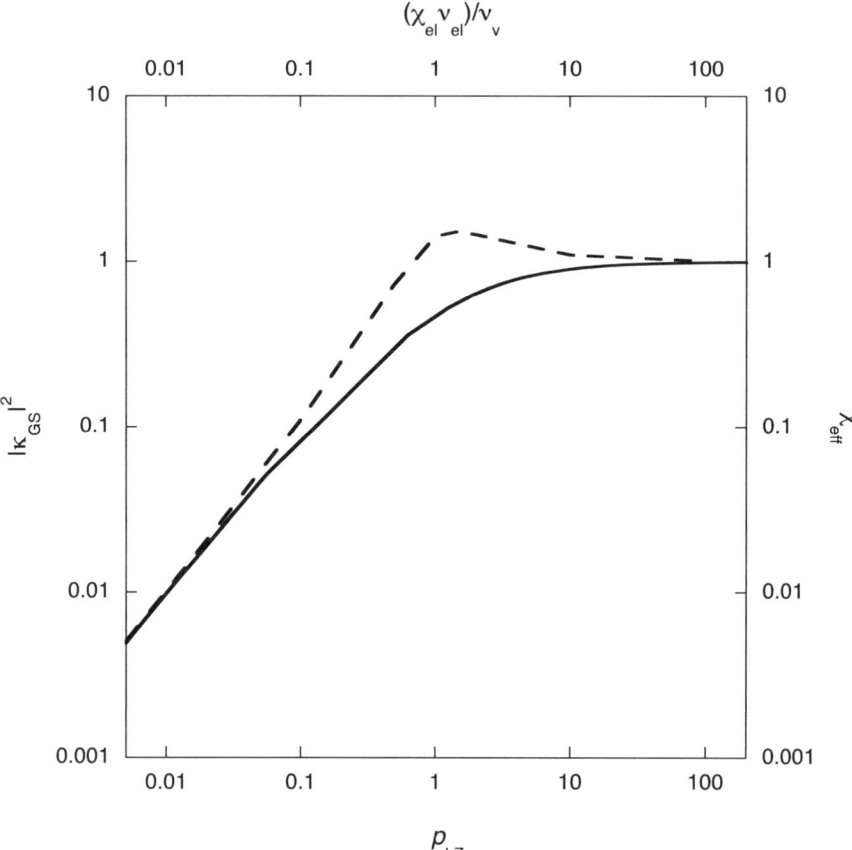

Figure 2.4 Electronic nonadiabatic factor (χ_{eff}) as a function of the ratio between nuclear and effective electronic frequencies $\nu_v/(\chi_{el}\nu_{el})$, or adiabaticity parameter ($|\kappa_{GS}|^2$) as a function of the Landau-Zener classical transition probability (p_{LZ}), illustrating their rapid convergence to the same asymptotic limits.

properties of the medium separating the donor from the acceptor. This dependence has been thoroughly investigated for a wide range of systems and can be expressed as electron tunnelling following the equation

$$\chi_{el} = \exp(-\beta_{el}r_{DA}) \tag{2.44}$$

where β_{el} is a tunnelling decay coefficient. The value of β_{el} depends on the properties of the medium and on the energy of the electron in the donor. Typical values are: $\beta_{el} \approx 1.6$ Å^{-1} for aqueous glasses,[53] $\beta_{el} \approx 1.3$ Å^{-1} for glycerol[44,54] and $\beta_{el} \approx 1.4$ Å^{-1} for proteins.[55]

2.2.4 Concerted Proton-Electron Transfers

The reaction barrier of CPET has additive contributions from the various molecular distortions occurring along the reaction path. In the case where the same bonds are involved in the synchronous electron *and* proton transfer, the contributions to the reaction barrier should be reasonably well described by the $\Delta V_{ad}^{\ddagger}$ term of eqn (2.2). This is trivially the case for HAT, but it will also be approximately the case of concerted proton–electron transfers involving the same bonds but different active orbitals for PT and ET, because most of the nuclear reorganization energies are already included in the reaction coordinate. Is this case it is still possible to consider additional contributions due to changes in geometry/hybridizations, but they are expected to be relatively minor contribution when compared with $\Delta V_{ad}^{\ddagger}$ calculated for the HAT. When the proton and the electron are synchronously transferred between different molecular groups, the additional contribution associated with the reorganization of bonds that are not present in the PT coordinate must be added to $\Delta V_{ad}^{\ddagger}$. This additional term, $\Delta G_{et}^{\ddagger}$, can be calculated using the ET theories discussed above.

When the proton and electron transfer involve different sites of the reactants, we may expect, in addition to the extra reorganization energy, that the electron transfer may be strongly non-adiabatic ($\chi_{el} \ll 1$) and substantially reduce the effective reaction frequency. Georgievskii and Stuchebrukhov addressed the transition between electronically adiabatic and non-adiabatic CPET and showed that the adiabatic factor that reflects the deviation from adiabaticity has the form[56]

$$\kappa_{GS} = \sqrt{2\pi p}\, \frac{e^{p \ln p - p}}{\Gamma(p+1)} \qquad (2.45)$$

where $\Gamma(x)$ is the gamma-function and p is the proton adiabaticity parameter, which is approximately the ratio of the proton tunnelling time τ_p and the electronic transition time τ_e: $p = \tau_p/\tau_e$. This adiabaticity parameter is to be compared with the amplitude (instead of the probability) of the usual Landau–Zener transition. Figure 2.4 shows the dependence of $|\kappa_{GS}|^2$ on $p_{LZ} = 2\pi p_{GS}$. The asymptotic limits are the same as those of χ_{el} in eqn (2.42) and only a small difference between $|\kappa_{GS}|^2$ and χ_{el} is observed in the transition between non-adiabatic and adiabatic systems. This could be anticipated from the reciprocal dependence between frequencies and tunnelling of transition times.

The theoretical treatments of nonadiabatic transitions and the formulation of hydrogen atom or proton tunnelling, as particular cases of such transitions, motivated various theoretical developments of CPET. In the simplest case, when only one initial and one final vibrational states are involved in the reaction, the CPET rate has been written in a form similar to that given by the nonadiabatic ET theory[56]

$$k_{CPET} = \frac{2\pi}{\hbar} |\kappa_{GS} V_{DA}{}^{(ad)}|^2 \frac{1}{\sqrt{4\pi\lambda k_B T}} \exp\left(-\frac{\Delta G^{\ddagger}}{k_B T}\right) \qquad (2.46)$$

where $V_{DA}^{(ad)}$ is the adiabatic vibronic coupling between the reactant and product vibronic states, and κ_{GS} accounts for the transition from adiabatic ($\kappa_{GS} \to 1$) to non-adiabatic ($\kappa_{GS} \to (2\pi p_{GS})^{1/2}$) regimes. This approach has been particularly proficuous,[57] but here we will follow an alternative path based on transition-state theory.

The transition between adiabatic PT and electronically non-adiabatic CPET can also be incorporated in the adiabatic PT rate constant using the electronic nonadiabatic factor χ_{el}, that depends on $(\chi_{el}v_{el})/v_v$ in the same way as $|\kappa_{GS}|^2$ depends on p_{LZ}. Additionally, we include in the energy barrier the contributions from the reorganizations of the parts of the molecular system involved in the synchronous proton and electron transfer, and express the rate constant for nonadiabatic CPET as

$$k_{CPET} = \chi_{eff} \nu_n \exp\left[-\frac{\Delta V_{ad}^{\ddagger} + \Delta G_{et}^{\ddagger}}{RT}\right] \qquad (2.47)$$

where $\nu_n = \kappa_{tun} K_c c\bar{\nu}_{e(AB)}$. The term χ_{eff} accounts for the electronic adiabaticity of the reaction (not to be confused with the vibrational adiabaticity, that describes the conservation of the zero-point energy along the reaction path). Typically CPETs become electronically non-adiabatic when the electron is transferred over a distance of more than 2 non-interacting chemical bonds, such as localized single bonds, or when the donor orbital is orthogonal with the acceptor orbital. The calculation of $\Delta V_{ad}^{\ddagger}$ and $\Delta G_{et}^{\ddagger}$ can be made as in eqn (2.19) and (2.39), which are both rooted in ISM and measure the number of systems in thermal equilibrium with sufficient energy for both proton and electron transfer.

2.3 Applications

2.3.1 HAT in the Benzyl/Toluene Self-Exchange

The textbook examples to distinguish between HAT and CPET are the self-exchanges in benzyl/toluene and phenoxyl/phenol.[58,59] The proton and the electron in the $PhCH_2^{\bullet} + PhCH_3$ self-exchange move synchronously from the benzylic carbon atom in one reactant to the analogous carbon atom in other reactant, and this is a typical HAT. On the other hand, in the $PhO^{\bullet} + PhOH$ self-exchange the proton is transferred between oxygen σ orbitals, whereas the electron is transferred between oxygen π orbitals. Transition-state theory with the vibrationally-adiabatic barrier calculated by established *ab initio* or semiempirical methods should be able to calculate the rate of the HAT in the benzyl/toluene system.

Before we compare the experimental and semiempirical HAT rates of the benzyl/toluene system, it is important to assess the validity of the semiempirical calculations with a set of representative and accurate HAT barrier heights. Lynch and Truhlar showed that the systems $OH + CH_4 \to CH_3 + H_2O$, $H + HO \to O + H_2$ and $H + H_2S \to H_2 + HS$ are very representative of this type of reaction and benchmarked comparisons to these systems. Truhlar and co-workers also reported benchmark calculations for the barrier heights of five degenerate and nearly degenerate HATs between hydrocarbon fragments ($CH_3^{\bullet}$ + CH_4 or C_2H_6, $C_2H_5^{\bullet}$ + CH_4 or C_2H_6 and n-$C_3H_7^{\bullet}$ + n-C_3H_8),[60] that provide additional reliable data to validate the application of semiempirical methods. Additionally, we include the simplest and best-known system, H + H_2,[61] in this comparison. The classical (zero-point-exclusive) barrier heights calculated with ISM with the data in Table 2.1[4] are compared in Table 2.2 with the barrier heights of the most accurate *ab initio* calculations.[1] The impressive

Table 2.1 Bond lengths, bond dissociation energies, vibrational frequencies of the molecules and ionization potentials and electron affinities of the radicals employed in the calculation of the energy barriers of PCET reactions.[a]

	$l_{eq}/\mathring{A}$	D^0_{298} / $kcal\ mol^{-1}$	ω_e/cm^{-1}	I_P/eV	E_A/eV
H_2	0.74144	104.2	4161	13.598	0.75419
CH_4	1.0870	104.9	2917	9.843	0.08
CH_3CH_3	1.0940	101.1	2954	8.117	−0.26
$CH_3CH_2CH_3$	1.107	97.8	2887	7.37	−0.321
$CH_3C_6H_5$	1.111	89.8	2934	7.242	0.912
C_6H_6	1.101	113.1	3062	8.32	1.096
H_2O	0.9575	119.0	3657	13.017	1.8277
OH	0.96966	102.2	3737.76	13.618	1.4611
HOOH	0.95	88.2	3608	11.35	1.078
CH_3OH	0.9451	104.2	3681	10.720	1.57
C_6H_5OH	0.956	86.5	3650	8.56	2.253
H_2S	1.3356	91.2	2615	10.422	2.317
CH_3SH	1.340	87.3	2610	9.262	1.867
HCl	1.27455	103.2	2886	12.968	3.6144

[a]From ref. 4.

Table 2.2 Classical (electronic) barrier heights of H-atom transfer reactions.

	ISM			*Ab initio*
System	$\Delta V_{cl}^0/kcal\ mol^{-1}$	m	$\Delta V_{cl}^{\ddagger}/kcal\ mol^{-1}$	$\Delta V_{cl}^{\ddagger}/kcal\ mol^{-1}$
$OH + CH_4 \rightarrow H_2O + CH_3$	−15.2	1.456	6.4	6.7[a]
$O + H_2 \rightarrow OH + H$	2.7	1.241	11.9	13.1[a]
$H + SH_2 \rightarrow H_2 + SH$	−15.4	1.572	3.7	3.6[a]
$H + H_2 \rightarrow H_2 + H$	0	1	10.1	9.9[b]
$CH_3 + CH_4 \rightarrow CH_4 + CH_3$	0	1	16.8	17.5[c]
$C_2H_5 + C_2H_6 \rightarrow C_2H_6 + C_2H_5$	0	1	17.2	16.7[c]
$C_3H_7 + C_3H_8 \rightarrow C_3H_8 + C_3H_7$	0	1	16.8	16.0[c]
$CH_3 + C_2H_6 \rightarrow CH_4 + C_2H_5$	−3.7	1	15.2	15.4[c]

[a]From ref. 1.
[b]From ref. 61.
[c]From ref. 60.

agreement between ISM and benchmark barrier heights validates the use of this semiempirical method to calculate HAT rate constants.

The benzyl/toluene self-exchange is difficult to measure experimentally because the reactants and products are identical. Alternatively, Jackson and O'Neill measured the HAT between benzyl radical and *m*-deuteriotoluene in the 124–168 °C range and obtained the Arrhenius equation log $(k/(M^{-1} s^{-1})) = 10.5$–$19.9/\theta$,[62] whereas Franz and co-workers measured the HAT between 2-allylbenzyl radical and *p*-xylene between 130 and 190 °C and obtained log $(k/(M^{-1} s^{-1})) = 7.1$–$13.4/\theta)$, $\theta = 2.3RT$ in kcal mol^{-1}.[63] ISM calculations[64] give, per equivalent hydrogen atom and for $m = 1$, log $(k/(M^{-1} s^{-1})) = 9.4$–$15.2/\theta)$ in the same temperature range. This, again, demonstrates the validity of ISM to calculate HAT rates.

2.3.2 PCET in the Phenoxyl/Phenol Self-Exchange

Hydrogen bonding is an integral part of the reactivity of the phenoxyl/phenol system. Thus, it is important to assess the role of H-bonding in the reactivity of representative systems. Zavitsas recently selected reliable activation energies for four symmetrical systems where H-bonding is likely to occur, to eliminate the effect of the reaction enthalpy from the barrier height.[65] These systems are presented in Table 2.3 and will be used here to investigate the role of H-bonding in their reactivity. The estimates of the H-bond binding energies were taken from earlier work.[4,16,20]

The H-atom exchange rate constant of the $H^{18}O + HOH$ system was measured in the gas phase at 300 K and is 1.3×10^5 M^{-1} s^{-1}.[66] Using the data in Table 2.1, we calculate $k_{ISM} = 4.4 \times 10^4$ M^{-1} s^{-1} ($K_c = 1$ M^{-1}) with a H-bond binding energy $D_{0AC} = 3.5$ kcal mol^{-1}. The experimental and calculated activation energies are presented in Table 2.3. The calculated value is higher than the experimental one but this may be assigned to the substantial tunnelling present. At 300 K the tunnelling correction is 96 and decreases to 5.5 at 420 K. The closest experimental analogy to the $CH_3O + HOCH_3$ self-exchange is hydroxylic H-abstraction of $((CH_3)_3C)_3COH$ by the relatively stable *tert-*

Table 2.3 Symmetrical $X^{\bullet} + HX \rightarrow XH + {}^{\bullet}X$ systems and their activation energies.

			I S M		*Exp.*[a]
System	D_{0AC}[a]/ kcal mol^{-1}	*T/K*	*m*	E_a/kcal mol^{-1}	E_a/kcal mol^{-1}
$OH + H_2O \rightarrow H_2O + OH$	3.5	300–420	1.327	8.6	4.2
$CH_3O + HOCH_3 \rightarrow CH_3OH + OCH_3$	3.5	213–293	1.343	6.3	2.6
$CH_3S + HSCH_3 \rightarrow CH_3SH + SCH_3$	2	298–373	1.505	8.8	5.2
$Cl + HCl \rightarrow HCl + Cl$	1	300–400	1.773	7.8	5.4–6.6
$PhO + HOPh \rightarrow PhOH + OPh$	3	243–331	1.714	5.1	2.3[b]

[a]From data in ref. 62 except where noted.
[b]From ref. 66.

butoxy radical, $(CH_3)_3CO^{\bullet}$, which has a rate constant of 3×10^4 M^{-1} s^{-1} at 293 K in solution.[67] With the data in Table 2.1 and $D_{0AC} = 3.5$ kcal mol^{-1}, we calculate $k_{ISM} = 3 \times 10^6$ M^{-1} s^{-1} ($K_c = 1$ M^{-1}). The overestimation of the rate is certainly due to the use of $K_c = 1$ M^{-1}, because the infrared spectrum of the reaction mixture revealed that "only a *very* small proportion of the alcohol" was not H-bonded to the solvent and the reaction occurs only from the reactants that are not H-bonded to the solvent.[67] Table 2.3 compares the experimental and calculated activation energies for this system, which follow the same trend as the HO + HOH system: the tunnelling corrections decrease from 328 at 213 K to 19 at 293 K and mask the actual barrier height. The best experimental analogy to the $CH_3S + HSCH_3$ system is the H-atom abstraction of hexanethiol by the octanethiyl radical in hydrocarbon solvents, which has a rate constant of 3×10^4 M^{-1} s^{-1} at 298 K.[68] With the data in Table 2.1 and $D_{0AC} = 2$ kcal mol^{-1}, we calculate $k_{ISM} = 3 \times 10^6$ M^{-1} s^{-1} ($K_c = 1$ M^{-1}), and ignoring H-bonding ($D_{0AC} = 0$) we obtain $k_{ISM} = 8 \times 10^2$ M^{-1} s^{-1}. This difference in the calculations with $D_{0AC} = 0$ or 2 kcal mol^{-1} is almost exclusively due to the pre-exponential factor because tunnelling corrections are modest for this system ($\kappa_{tun} \leq 2$) and the activation energies are very similar. Better estimates of K_c are necessary to improve the accuracy of the rate calculations and the values presented here represent the upper and lower limits expected by ISM. Finally, the H-atom exchange in Cl + HCl in the gas phase at 358 K was measured as $(2.5 \pm 1.5) \times 10^6$ M^{-1} s^{-1},[69] and using the data in Table 2.1 we calculate $k_{ISM} = 2.2 \times 10^7$ M^{-1} s^{-1} ($K_c = 1$ M^{-1}) with $D_{0AC} = 1$ kcal mol^{-1} and $k_{ISM} = 1.7 \times 10^6$ M^{-1} s^{-1} with $D_{0AC} = 0$. The experimental activation energy is between 5.4 and 6.6 kcal mol^{-1},[65] and the calculated activation energy is 7.8 kcal mol^{-1}, with a negligible dependence on D_{0AC}.

The activation energies calculated for the temperature range of the experimental studies exceed the experimental activation energies by 1–4 kcal mol^{-1}. However, the calculated activation energies are based on rate constants calculated for H-bonded systems, that is, with barrier heights measured from the minimum of the H-bond energy to the maximum of the H-abstraction barrier. On the other hand, the uncertainty of the value of K_c makes absolute comparisons with the experimental rates very challenging. In view of these difficulties, the reactivity trends reflected by ISM are in good agreement with the experimental observations.

The closest experimental analogue for the PhO + HOPh system is the abstraction of the hydroxylic hydrogen of α-naphthol or β-naphthol by the phenoxyl radical in weakly polar solvents.[70] It was found that phenoxyl radical has a two orders of magnitude higher reactivity than the peroxyl radical and this was assigned to a difference in pre-exponential factor of the reaction initiated by the PhO$^{\bullet}$ radical or by the $(CH_3)_3COO^{\bullet}$ radical. ISM calculations with the data in Table 2.1 using HOOH as a model for $(CH_3)_3COOH$ give very similar activation energies for the PhO$^{\bullet}$ + OHPh and $(CH_3)_3COO^{\bullet}$ + OHPh reactions, and also impart the reactivity difference to the pre-exponential factors. The activation energy for the PhO$^{\bullet}$ + OHPh reaction is in line with the others discussed above, and very close to the activation energy of 5.6 kcal mol^{-1} calculated by Mayer and co-workers.[58] *Ab initio* calculations on the PhO$^{\bullet}$ + OHPh system showed the presence of a H-bonded complex, and the electronic energy barrier from this complex to the transition-state (C_2 symmetry) was reported as 8.6 kcal mol^{-1},[58] which should be compared with the ISM classical energy barrier of 8.0 kcal mol^{-1}. The consistency among these barriers gives further credence to the argument that it is the pre-exponential factor (*i.e.*, the value of K_c) that distinguishes the reactivity of phenoxyl radicals from that of peroxyl radicals. The fact that the electron accompanying this proton transfer is exchanged between orbitals of different symmetries does not seem to contribute meaningfully to the barrier because the reactive bonds are the same as in a HAT reaction.

2.3.3 CPET in Soybean Lipoxygenase-1

The formal H-atom abstraction of linoleic acid by soybean lipoxygenase-1 (SLO)

(9Z,12Z)-octadeca-9,12-dienoic acid
linoleic acid

(S,9Z,11E)-13-hydroperoxyoctadeca
-9,11-dienoic acid
13-HPOD

R_1 = R_2 =

pentane   octanoic acid

(VI)

is a CPET where the electron is transferred from the π-system of the linoleic acid (π_D) to the iron (Fe_A) and promotes the transfer of the proton from the C(11) carbon atom of linoleic acid to the oxygen atom of the Fe-bound OH ligand. We have addressed this system combining the proton transfer adiabatic coordinate with the distance-dependent electronic factor of the nonadiabatic electron transfer.[71] The PT coordinate was constructed with the O–H bond of H$_2$O and the C–H bond of benzene, because they typify the reactive bonds and give an adiabatic reaction energy of –5.5 kcal mol^{-1}, identical to the experimental ΔG^0.[72] The reaction frequency in eqn (2.47) is dominated by the low value of the electronic nonadiabatic factor (χ_{el}). Using reasonable estimates for the properties of the medium ($\beta_{el} \approx 1.4$ Å^{-1})[55] and for the distance between electron donor and acceptor ($r_{DA} \approx 4$ Å),[72] eqn (2.44) gives $\chi_{el} = 3.7 \times 10^{-3}$ s^{-1}. On the other hand the H-bond frequency and the barrier height for the proton transfer are $\bar{\nu}_{e(AB)} = 6.2 \times 10^{12}$ s^{-1} and $\Delta V_{ad}^{\ddagger} = 14.69$ kcal mol^{-1} at 25 °C. The calculation of the PCET rate constant additionally requires the contribution of ET to the iron complex, $\Delta G_{et}^{\ddagger}$, which is not accounted in the PT coordinate.

The contribution of ET to the reaction barrier was ignored in our earlier approach to this system, but is now calculated using eqn (2.40) using the experimental bond lengths ($l_{ox} = 1.53$ and $l_{red} = 0.96$ Å) and force constants ($f_{ox} = 366$ and $f_{red} = 230$ kcal mol^{-1} Å^{-2}) of Fe^{3+}–OH$_2$ and Fe^{2+}–OH$_2$ in Fe(H$_2$O)$_6^{3+/2+}$ and, for a self-exchange reaction,[11]

$$\Delta G_0^{\ddagger} = \frac{1}{2} \frac{f_{ox} + f_{red}}{2} \left(\frac{d}{2} \right)^2 \tag{2.48}$$

which gives $\Delta G_0^{\ddagger} = 7.2$ kcal mol^{-1}. This, however, was calculated as if the ET was isothermic. The simplest correction to account for the relatively small exothermicity of the reaction ($\Delta G^0 = -5.5$ kcal mol^{-1}), is that of the Marcus cross relation, eqn (2.27), which leads to a barrier height of 4.7 kcal mol^{-1}. This method to calculate the reaction barrier is only applicable to ET reactions between a donor and an acceptor. However, in this CPET the donor is the CH bond that is already included in the reaction coordinate, and the ET process should be regarded as having only one reactant. This is also true in the case of heterogenous ET between a redox couple in solution and a metal electrode. Marcus has shown that the barrier for heterogenous ET is simply half that of the barrier for the analogous homogenous ET,

$$\Delta G_{het}^* = \frac{1}{2} \Delta G_{hom}^* \tag{2.49}$$

because the response of the metal electrode to the transfer of the electron with the redox couple does not involve energy changes.[73] Within these approximations, we finally obtain $\Delta G_{et}^{\ddagger} = 2.35$ kcal mol^{-1}. The CPET with $\Delta V_{ad}^{\ddagger} = 14.69$ kcal mol^{-1}, $\Delta G_{et}^{\ddagger} = 2.35$ kcal mol^{-1}, $\chi_{eff} = 0.063$, $c\bar{\nu}_{e(AB)} = 6.2 \times 10^{12}$ s^{-1} and a tunnelling correction of $\kappa_{tun} = 106$ at 298 K, is $k_{CPET} = 13$ s^{-1}. The experimental rate constant is 330 s^{-1}.[74] Although this can be regarded as a rather eclectic approach, the results are insightful and in qualitative agreement with the experiment.

2.4 Conclusions

Concerted proton–electron transfers combine reactivity factors from hydrogen atom, proton and electron transfers that are difficult to reconcile because they are rooted in opposing approximations. In the transition-state view of reactivity, the system evolves adiabatically from reactants to products in the same potential energy surface and transition-state properties, such as the transition-state energy and partition function, determine the rate constant. Radiationless transition theories describe the evolution of systems between weakly interacting potential energy surfaces and relate the reactivity to the magnitudes of the Franck–Condon factor and of the electronic coupling. We briefly reviewed these concepts and rather than following the more common view of PCET as radiationless transitions, we explored a formulation that emphasises the reactivity factors of transition-state theory. Our view offers a clear relation between reactivity and molecular parameters, such as bond lengths, force constants, bond strengths, ionization potentials, electronic affinities, H-bond binding energies and reaction energies. This picture was completed with a quantification of nonadiabaticity and its dependence on the frequency factors of transition-state and radiationless transition theories.

References

1. B. J. Lynch and D. G. Truhlar, *J. Phys. Chem. A*, 2003, **107**, 8996–8999.
2. *Hydrogen-Transfer Reactions*, ed. J. T. Hynes, J. P. Klinman, H. H. Limbach and R. L. Schowen, Wiley-VCH, Weinheim, Germany, 2007.
3. L. G. Arnaut, S. J. Formosinho and H. D. Burrows, *Chemical Kinetics*, Elsevier, Amsterdam, 2007.
4. L. G. Arnaut and S. J. Formosinho, *Chem.–Eur. J.*, 2008, **14**, 6578–6587.
5. M. H. V. Huynh and T. J. Meyer, *Chem. Rev.*, 2007, **107**, 5004–5064.
6. T. J. Meyer, M. H. V. Huynh and H. H. Thorp, *Angew. Chem., Int. Ed.*, 2007, **46**, 5284–5304.
7. J. L. Dempsey, J. R. Winkler and H. B. Gray, *Chem. Rev.*, 2010, **110**, 7024–7039.
8. Z. D. Nagel and J. P. Klinman, *Chem. Rev.*, 2010, **110**, PR41–PR67.
9. C. Costentin, M. Robert and J.-M. Savéant, *Chem. Rev.*, 2010, **110**, PR1–PR40.
10. J. M. Mayer, *Annu. Rev. Phys. Chem.*, 2004, **55**, 363–390.
11. S. J. Formosinho, L. G. Arnaut and R. Fausto, *Prog. Reaction Kinetics*, 1998, **23**, 1–90.
12. L. G. Arnaut, A. A. C. C. Pais, S. J. Formosinho and M. Barroso, *J. Am. Chem. Soc.*, 2003, **125**, 5236–5246.
13. B. C. Garrett and D. G. Truhlar, *J. Phys. Chem.*, 1979, **83**, 2921–2926.
14. H. S. Johnston and C. Parr, *J. Am. Chem. Soc.*, 1963, **85**, 2544–2551.
15. R. G. Parr, L. V. Szentpály and S. Liu, *J. Am. Chem. Soc.*, 1999, **121**, 1922–1924.
16. M. Barroso, L. G. Arnaut and S. J. Formosinho, *ChemPhysChem*, 2005, **6**, 363–371.
17. L. Pauling, *J. Am. Chem. Soc.*, 1947, **69**, 542–553.
18. A. J. C. Varandas, F. B. Brown, C. A. Mead, D. G. Truhlar and B. C. Garrett, *J. Chem. Phys.*, 1987, **86**, 6258–6269.
19. E. R. Lippincott and R. Schroeder, *J. Chem. Phys.*, 1955, **23**, 1099–1106.
20. M. Barroso, L. G. Arnaut and S. J. Formosinho, *J. Phys. Chem. A*, 2007, **111**, 591–602.
21. N. Stahl and W. P. Jencks, *J. Am. Chem. Soc.*, 1986, **108**, 4196–4205.
22. G. Litwinienko and K. U. Ingold, *Acc. Chem. Res.*, 2007, **40**, 222–230.
23. D. W. Snelgrove, J. Lusztyk, J. T. Banks, P. Mulder and K. U. Ingold, *J. Am. Chem. Soc.*, 2001, **123**, 469–477.

24. S. J. Formosinho, *J. Chem. Soc., Faraday Trans.* 2, 1976, 72, 1313–1331.
25. J. Jortner and J. Ulstrup, *Chem. Phys. Lett.*, 1979, **63**, 236–239.
26. S. J. Formosinho and L. G. Arnaut, *Adv. Photochem.*, 1991, **16**, 67–117.
27. R. E. Galian, G. Litwinienko, J. Pérez-Prieto and K. U. Ingold, *J. Am. Chem. Soc.*, 2007, **129**, 9280–9281.
28. J. Ulstrup, *Charge Transfer Processes in Condensed Media*, Spinger-Verlag, Berlin, 1979.
29. R. A. Marcus, *J. Chem. Phys.*, 1956, **24**, 966–978.
30. R. A. Marcus, *J. Chem. Phys.*, 1957, **26**, 867–871.
31. R. A. Marcus, *J. Chem. Phys.*, 1957, **26**, 872–877.
32. P. George and J. S. Griffith, in *The Enzymes*, ed. P. D. Boyer, H. Lardy and K. Myrbäck, Academic Press, New York, 1959, vol. 1, p. 347.
33. R. A. Marcus, *Faraday Discuss. Chem. Soc.*, 1960, **29**, 21–31.
34. S. F. Nelsen, M. N. Weaver, J. R. Pladziewicz, L. K. Ausman, T. L. Jentzsch and J. J. O'Konek, *J. Phys. Chem. A*, 2006, **110**, 11665–11676.
35. C. Serpa, P. J. S. Gomes, L. G. Arnaut, S. J. Formosinho, J. Seixas de Melo and J. Pina, *Chem.–Eur. J.*, 2006, **12**, 5014–5023.
36. H. Miyasaka, S. Kotani, A. Itaya, G. Schweitzer, F. C. De Schryver and N. Mataga, *J. Phys. Chem. B*, 1997, **101**, 7978–7984.
37. J. R. Miller, L. T. Calcaterra and G. L. Closs, *J. Am. Chem. Soc.*, 1984, **106**, 3047–3049.
38. J. R. Miller, J. V. Beitz and R. K. Huddleston, *J. Am. Chem. Soc.*, 1984, **106**, 5057–5068.
39. G. L. Closs, L. T. Calcaterra, N. J. Green, K. W. Penfield and J. R. Miller, *J. Phys. Chem.*, 1986, **90**, 3673–3683.
40. N. Mataga, T. Asahi, Y. Kanda, T. Okada and T. Kakitani, *Chem. Phys.*, 1988, **127**, 249–261.
41. I. R. Gould, R. Moody and S. Farid, *J. Am. Chem. Soc.*, 1988, **110**, 7242–7244.
42. D. M. Guldi and K.-D. Asmus, *J. Am. Chem. Soc.*, 1997, **119**, 5744–5745.
43. C. Serpa and L. G. Arnaut, *J. Phys. Chem. A*, 2000, **104**, 11075–11086.
44. P. J. S. Gomes, C. Serpa, R. M. D. Nunes, L. G. Arnaut and S. J. Formosinho, *J. Phys. Chem. A*, 2010, **114**, 2778–2787.
45. R. P. Van Duyne and S. F. Fischer, *Chem. Phys.*, 1974, **5**, 183–197.
46. N. R. Kestner, J. Logan and J. Jortner, *J. Phys. Chem.*, 1974, **78**, 2148–2166.
47. S. Efrima and M. Bixon, *Chem. Phys. Lett.*, 1974, **25**, 34–37.
48. M. Bixon and J. Jortner, *J. Phys. Chem.*, 1991, **95**, 1941–1944.
49. C. Serpa, L. G. Arnaut, S. J. Formosinho and K. R. Naqvi, *Photochem. Photobiol Sci.*, 2003, **2**, 616–623.
50. J. Jortner, *J. Chem. Phys.*, 1976, **64**, 4860–4867.
51. D. DeVault, *Quart. Rev. Biophys.*, 1980, **13**, 387–564.
52. R. A. Marcus and N. Sutin, *Biochim. Biophys. Acta*, 1985, **811**, 265–322.
53. O. S. Wenger, B. S. Leigh, R. M. Villahermosa, H. B. Gray and J. R. Winkler, *Science*, 2005, **307**, 99–102.

54. P. J. S. Gomes, C. Serpa, R. M. D. Nunes, L. G. Arnaut and S. J. For-mosinho, *J. Phys. Chem. A*, 2010, **114**, 10759–10760.
55. C. C. Moser, J. M. Keske, K. Warncke, R. S. Farid and P. L. Dutton, *Nature*, 1992, **355**, 796–802.
56. Y. Georgievskii and A. A. Stuchebrukhov, *J. Chem Phys.*, 2000, **113**, 10438–10450.
57. S. Hammes-Schiffer and A. V. Soudackov, *Chem. Rev.*, 2010, **110**, 6939–6960.
58. J. M. Mayer, D. A. Hrovat, J. L. Thomas and W. T. Borden, *J. Am. Chem. Soc.*, 2002, **124**, 11142–11147.
59. J. H. Skone, A. V. Soudackov and S. Hammes-Schiffer, *J. Am. Chem. Soc.*, 2006, **128**, 16655–16663.
60. A. Dybala-Defratyka, P. Paneth, J. Pu and D. G. Truhlar, *J. Am. Chem. Soc.*, 2004, **108**, 2475–2486.
61. B. G. Johnson, C. A. Gonzales, P. M. W. Gill and J. A. Pople, *Chem. Phys. Lett.*, 1994, **221**, 100–108.
62. R. A. Jackson and D. W. O'Neill, *J. Chem. Soc., Chem. Commun.*, 1969, 1210–1211.
63. J. A. Franz, M. S. Alnajjar, R. D. Barrows, D. L. Kaisaki, D. M. Camaioni and N. K. Suleman, *J. Org. Chem.*, 1986, **51**, 1446–1456.
64. L. G. Arnaut, M. Barroso, D. Oliveira, *Intersecting/Interacting State Model*, 2006, http://www.ism.qui.uc.pt:8180/ism/
65. A. A. Zavitsas, *J. Phys. Chem. A*, 2010, **114**, 5113–5118.
66. M. K. Dubey, R. Mohrschladt, N. M. Donahue and J. G. Anderson, *J. Phys. Chem. A*, 1997, **101**, 1494–1500.
67. D. Griller and K. U. Ingold, *J. Am. Chem. Soc.*, 1974, **96**, 630–632.
68. M. S. Alnajjar, M. S. Garrossian, S. T. Autrey, K. F. Ferris and J. A. Franz, *J. Phys. Chem.*, 1992, **96**, 7037–7043.
69. M. Kneba and J. Wolfrum, *J. Phys. Chem.*, 1979, **83**, 69–73.
70. M. Foti, K. U. Ingold and J. Lusztyk, *J. Am. Chem. Soc.*, 1994, **116**, 9440–9447.
71. M. Barroso, L. G. Arnaut and S. J. Formosinho, *J. Phys. Org. Chem.*, 2008, **21**, 659–665.
72. E. Hatcher, A. V. Soudackov and S. Hammes-Schiffer, *J. Am. Chem. Soc.*, 2004, **126**, 5763–5775.
73. R. A. Marcus, *Annu. Rev. Phys. Chem.*, 1964, **15**, 155–196.
74. M. J. Knapp, K. Rickert and J. P. Klinman, *J. Am. Chem. Soc.*, 2002, **124**, 3865–3874.

CHAPTER 3

Experimental Approaches Towards Proton-Coupled Electron Transfer Reactions in Biological Redox Systems

SIBYLLE BRENNER,[†] SAM HAY, DERREN J. HEYES AND NIGEL S. SCRUTTON*

Manchester Interdisciplinary Biocentre and Faculty of Life Science, University of Manchester, Manchester, UK

3.1 Introduction

3.1.1 Definitions

Proton-coupled electron transfer (PCET) reactions play a fundamental role in the respiratory chain and in photosynthesis.[1–8] In both membrane-bound systems, an electrochemical gradient is built up across the lipid bilayer by separating protons from electrons; the resulting chemiosmotic proton potential serves to fuel a proton-driven pump synthesising the universal biological energy equivalent ATP (adenosine triphosphate). In addition to the relevance of PCET in these energy conversions, the coupled transfer of electrons and protons is an

[†]Current address: Department of Anatomy and Structural Biology and Gruss-Lipper Biophotonics Center, Albert Einstein College of Medicine, New York, USA

RSC Catalysis Series No. 8
Proton-Coupled Electron Transfer: A Carrefour of Chemical Reactivity Traditions
Edited by Sebastião Formosinho and Mónica Barroso
© Royal Society of Chemistry 2012
Published by the Royal Society of Chemistry, www.rsc.org

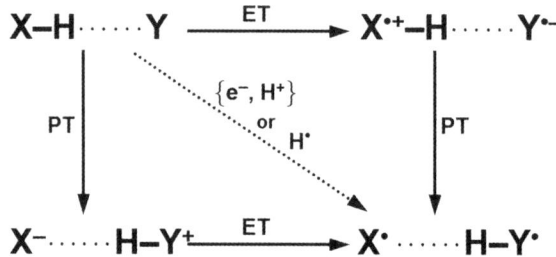

Figure 3.1 Square scheme illustrating stepwise and concerted PCET reactions. The borders reflect stepwise PCET involving real chemical intermediates, while the entire inside plane comprises concerted PCET mechanisms with a single transition state. The diagonal represents the simultaneous transfer of an electron and a proton. If the charges are not orbitally separated, this elementary step is a hydrogen atom transfer (HAT) symbolised by H$^\bullet$. PT, proton transfer; ET, electron transfer. See text for further information. Adapted from ref. 21.

essential mechanistic feature of several metabolic redox enzymes, which catalyse a broad variety of reactions ranging from the formation of building blocks for biosynthetic pathways, as found in ribonucleotide reductase, to detoxification reactions, such as those catalysed by superoxide dismutases.[9–17]

A complicating feature in discussing PCET mechanisms is posed by the inconsistency in the definitions and nomenclatures used in different publications. Generally, the borders of a square scheme (Figure 3.1) are employed to illustrate stepwise mechanisms (see *e.g.* ref. 18–21), in which an elementary proton transfer (PT) step is followed by an electron transfer (ET) and *vice versa*. While these sequential reactions include the formation of real chemical intermediates, a concerted transfer of the proton and the electron represents an elementary reaction with a single transition state (TS) and is illustrated by the interior plane of the square (Figure 3.1; ref. 20 and 21). The distance from the diagonal reflects the degree of charge separation during the reaction step. Thus, the TS of a concerted PCET can either resemble the protonated, oxidised species or the deprotonated, reduced intermediate formed in a corresponding stepwise mechanism. Some publications define PCET exclusively as the inside of the thermodynamic cycle[18,20,21] reasoning that the stepwise mechanisms could be described by two separate rate constants for PT and ET (k_{PT}, k_{ET}) using the conventional formalism for kinetically linked reaction steps.[21] In contrast, the term PCET can also be found to incorporate both stepwise and concerted mechanisms.[4,22–24] In this review, experimental approaches to study PCET reactions will be introduced from a kineticist's viewpoint, and the complications in kinetically distinguishing between stepwise and concerted mechanisms will become evident. Therefore, this review will use a broader definition of PCET including both concerted and kinetically linked stepwise mechanisms.

PCET reactions can also be distinguished by the nature of the proton and electron acceptors. In a uni-directional PCET reaction, the proton and the

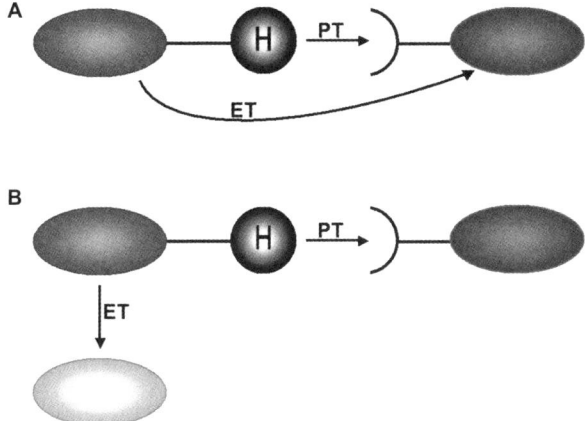

Figure 3.2   Uni-directional (A) and bi-directional (B) PCET. Synonyms for uni-directional PCET are collinear PCET,[20] concerted electron-proton transfer (CEP, ref. 236), electron–proton transfer (EPT, 120 and 237), concerted proton–electron transfer (CPET, 238 and 239) and concerted electron transfer proton transfer (ETPT, ref. 240). Bi-directional PCET is also termed orthogonal PCET,[20] bi-directional concerted electron–proton transfer (CEP, ref. 236) and multisite electron proton transfer (MS-EPT, ref. 237). Adapted from ref. 21.

electron are transferred to the same acceptor, whereas a bi-directional mechanism involves spatially separated acceptors for the proton and the electron (Figure 3.2; ref. 20 and 21). Various synonymous expressions are in use, some of which are listed in the figure legend of Figure 3.2, illustrating the ambiguous nomenclature in the field of PCET. A special case of PCET is given by the diagonal of the square scheme (Figure 3.1), which represents the simultaneous transfer of a proton and an electron. If both charges originate from and are transferred to the same orbital, both proton and electron cover the same spatial distance, the PCET is uni-directional and the reaction is called hydrogen atom transfer (HAT).[25] Alternatively, the diagonal can stand for kinetically simultaneous, yet orbitally separated PCET events, which cannot be defined as HAT reactions. In fact, the term PCET was originally coined to distinguish between these simultaneous non-HAT reactions and real HAT,[26–28] *i.e.* PCET was merely used to describe one version of the diagonal path in the square scheme (Figure 3.1). While most publications include HAT as a special case of concerted PCET reactions, HAT has been factored out in an electrochemical context, where the electrode serves as the electron donor/acceptor and acids/bases as the proton donors/acceptors.[29]

In a PCET, the electron does not have to be coupled to a particular proton throughout the entire reaction.[20] A prominent example is the class I ribonucleotide reductase from *E. coli*, which catalyses the reduction of nucleoside diphosphates (NDPs) to deoxynucleotide diphosphates (dNDPs),[30] the building blocks of deoxyribonucleic acid (DNA). An electron is transferred from a

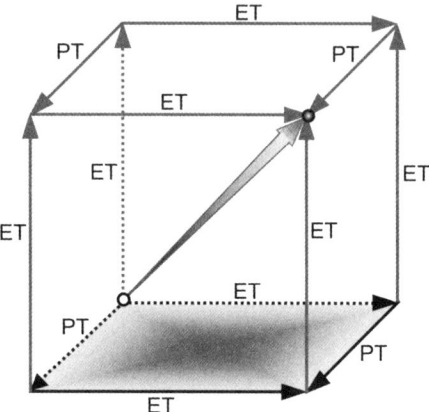

Figure 3.3 Cubic scheme illustrating the net transfer of one proton and two electrons. The initial state is indicated by the open circle in the lower-left rear corner. The bottom of the cube shaded in grey tones is analogous to the square scheme in Figure 3.1. The body diagonal connecting the initial state (open circle) with the final state (closed circle) reflects the simultaneous transfer of one proton and two electrons. If the three charges are not orbitally separated throughout the transfer, then this elementary reaction reflects a hydride ion transfer (HIT). Please note that in those cases where the total number of transferred charges exceeds three, the schematic representation is obviously not as straightforward. ET, electron transfer; PT, proton transfer.

diferric tyrosyl cofactor in the β-subunit through a series of a tryptophanyl and several tyrosyl radicals to a cysteine residue, which is located 35 Å away in the α-subunit.[9,20,21,31–35] This ET is accompanied by one uni-directional and several bi-directional short-distance PT steps.[20,21] The mechanism of ribonucleotide reductase indicates that the stoichiometry of PCET reactions is not restricted to one electron and one proton. In cytochrome *c* oxidase, one electron and two protons are involved in a PCET mechanism, whereby one proton is used for the formation of the product H_2O and the second proton is pumped across the membrane yielding a proton-gradient that fuels ATP-synthase.[36–43] If the coupled transfer of one electron and two protons can be classified as PCET, the transfer of two electrons and one proton – a net hydride transfer – is also implicit in the generalised definition of PCET.[18] Graphically, hydride ion transfer can be depicted by extending the square scheme in Figure 3.1 to a cube (Figure 3.3), in which the body diagonal represents the simultaneous transfer of two electrons and one proton. Treating hydride transfer as a special case of PCET reactions immediately raises the question of whether a net hydride transfer always occurs as a single chemical entity (hydride ion transfer, HIT) or whether different scenarios such as stepwise mechanisms are possible.

In biological systems, the redox coenzymes NADH (reduced nicotinamide adenine dinucleotide phosphate) and NADPH, the 2′-phosphoric acid derivative of NADH, belong to the most prominent and important examples of hydride donors. More than 400 redox reactions involve the interconversion of nicotinamide cofactors.[44] Furthermore, NADH plays a crucial role in the mitochondrial respiration chain by providing two reducing equivalents for the reduction of molecular oxygen to water.[44] In recent chemical publications, the actual mechanism of the observed net hydride transfer has been scrutinised using NADH analogues.[44–46] Challenging the one-step HIT mechanism, two potential multi-step hydride transfer mechanisms have been suggested, namely ET followed by HAT and ET–PT–ET.[44] Both stepwise reaction sequences are expected to yield transient radical species of the nicotinamide cofactor, the detection of which could serve as evidence against HIT. With the exception of the cited publications,[44–46] little enzymatic or non-enzymatic evidence has so far been provided to substantiate a stepwise mechanism due to the elusive nature of the transiently formed radicals.[44] Further advances in the theoretical and experimental analysis of net hydride transfer reactions will have to elucidate whether stepwise mechanisms do compete with one-step hydride transfer reactions (see Section 3.2).

3.1.2   Thermodynamics of PCET Reactions

The square scheme in Figure 3.1 represents a thermodynamic cycle for the stepwise transfer of an electron and a proton in any order, in which the diagonal path is the sum of the sequential reaction steps, thus combining redox with acid–base chemistry.[18,19,47] Any change in the pK_a value has to be balanced by a correspondingly changed redox potential and *vice versa*.[18,19] This E–pK_a compensation, also known as the redox-Bohr effect,[48] states that more protonated species are stronger oxidants, *i.e.* have a more positive redox potential, than their deprotonated counterparts. Likewise, the pK_a value of an acid–base system depends on the redox state of the molecule, *i.e.* a molecule is more acidic (has a lower pK_a value) in the oxidised form. This thermodynamic link between redox and acid–base properties is reflected in the Nernst equation, which predicts a change in redox potential at 298 K of 0.059 V $\times p/e$ per pH unit, where p and e are the number of transferred protons and electrons, respectively (see below; ref. 18 and 19). Based on the Nernst equation, eqn (3.1) expresses the thermodynamic relationship between pK_a values and redox potentials (E):[18]

$$\begin{aligned}\Delta G(H^+/e^-) &= 2.3RT\, pK_a(XH) + FE(X^\bullet/X^-) \\ &= FE(HX^{+\bullet}/HX) + 2.3RT\, pK_a(XH^{+\bullet})\end{aligned} \tag{3.1}$$

where ΔG is the Gibb's free reaction energy for the transfer of one proton and one electron, R the gas constant, F the Faraday constant and T the absolute temperature. The factor 2.3 arises from the conversion of a mathematical expression using the natural logarithm (ln) into the logarithm with the base 10 (log) (ln 10 = 2.3). The redox couples ($HX^{+\bullet}/HX$, $X^\bullet/X^-$) and acids (XH, $XH^{+\bullet}$) correspond to the corners in Figure 3.1.

The redox-Bohr effect and, thus, the pH dependence of redox potentials can largely be explained by the effect of electrostatics on the redox potentials.[25] Using the principles of a thermodynamic cycle, the formal redox potential ($E^{\circ\prime}$)[49] of any couple $Ox^{n+} + e^- \rightarrow Red^{(n-1)+}$ can be calculated from the gas phase ionisation energy for $Red^{(n-1)+}$ (I_{Red}) and the solvation free energies of the ions Ox^{n+} and $Red^{(n-1)+}$ (ΔG_{solv_Ox} and ΔG_{solv_Red}). An additional constant parameter (C) relates these values to a reference ion with zero standard enthalpy and Gibb's free energy of formation, which is conventionally chosen to be the hydrogen ion.[50] Therefore, the equation expressing the formal redox potential (eqn (3.2)) also includes the bond dissociation energy for dihydrogen (H_2; $\Delta G^{\circ\prime}$), the H-atom ionisation energy (I_H) and the free solvation energy of a proton ($\Delta G^{\circ\prime}_{solv\_H}$).

$$E^{\circ\prime} = \Delta G_{solv\_Ox} - \Delta G_{solv\_Red} + I_{Red} + C$$
$$\text{where } C = -0.5\Delta G^{\circ\prime}(H_2) - I_H - \Delta G^{\circ\prime}_{solv\_H} \tag{3.2}$$

The consequence of eqn (3.2) is an increase in $E^{\circ\prime}$ upon accumulation of positive charge, as expressed by the redox-Bohr effect (see above): a more protonated species is a stronger oxidant. Similarly, the redox potential difference between two adjacent redox couples $Ox^{(n+1)+}/Red^{n+}$ and $Ox^{n+}/Red^{(n-1)+}$ is mainly a result of the different charges.

The thermodynamic effects of a PCET can be rationalised by eqn (3.2). In a concerted transfer of the oppositely charged electron and proton, the charges of electron/proton donor and electron/proton acceptor are unaffected, *e.g.* Ox_1^{n+}/Red_1^{n+} and Ox_2^{n+}/Red_2^{n+}. Therefore, the electrostatic contribution to the redox potential difference $\Delta E^{\circ\prime}$ is negligible and the redox potentials $E^{\circ\prime}_1$ and $E^{\circ\prime}_2$ are comparatively close. This effect is known as redox potential levelling (see examples below; ref. 25). The thermodynamic advantage of a concerted *versus* a stepwise PCET can be illustrated using the square scheme for PCET reactions (Figure 3.1): the concerted transfer of the opposite charges circumvents the formation of potentially high-energy intermediates (the lower-left and upper-right corners of the cycle), as the charge separation may be energetically unfavourable (*cf.* eqn (3.2)). Thus, by combining an endergonic reaction step with an exergonic event in one elementary reaction, a thermodynamically favourable or less unfavourable reaction may be achieved.[18,19]

Redox potential levelling plays an important role in biological redox reactions involving multiple oxidation states of a single redox cofactor, since a series of ET steps can occur over a narrow, biologically accessible potential range.[25] Enzymes including the flavin cofactors FAD (flavin adenine dinucleotide) and/or FMN (flavin mononucleotide) provide an especially versatile redox and acid–base chemistry resulting in closely spaced redox potentials, whose relative positions can be fine-tuned by both the protonation state of the one- and two-electron reduced cofactor and the protein environment.[51–61]

Closely spaced redox potentials can also be found in transition metal cofactors, even if the overall charge does vary between the various oxidation states. This feature directly results from the small and similar ionisation energies for the 3d, 4d and 5d electron shells, which is due to the sufficient shielding of the electrons from the positively charged nucleus.[25] A classical example of biological significance is the manganese cluster found in the oxygen evolving complex of photosystem II.[62–71]

Experimentally, redox potentials can be determined using optical redox potentiometry given that the UV-vis signal of the redox couple changes in relation to its oxidation state. Using chemical reductants such as dithionite, optical spectra are collected for defined solution potentials ranging from the completely oxidised to the fully reduced sample. The resulting absorbance changes can then be plotted *versus* the respective potential and analysed by a modified Nernst equation yielding redox midpoint potentials characteristic for the experimental pH, temperature and buffer system.[53,72] Conducting redox titrations at various pH values can be used in so-called Pourbaix[73] or potential/pH diagrams to determine the ratio of p/e from the slope (see above; ref. 19). Kinks found in Pourbaix diagrams correspond to the pK_a values of involved acid–base couples.[25,29] The experimental determination of pK_a values from the pH-dependent redox potentials may, however, be impeded by the limited pH stability of the investigated enzyme.[72] Optical redox potentiometry has been widely applied in biological systems such as enzymes with heme or flavin cofactors.[53–55,72,74–76] For enzymatic systems lacking any redox-dependent UV-vis signal, electron paramagnetic resonance (EPR; see *e.g.* ref. 77–79) or cyclic protein film voltammetry (PFV; ref. 80) can be employed to determine the redox potentials. In PFV, a protein monolayer is absorbed onto an electrode and redox potentials can be obtained by measuring the current upon slowly tuning the electrode potential in the absence of substrate.[80] At low scan rates, PFV functions as an equilibrium technique and the redox states of the active sites reflect the ratios predicted by the Nernst equation.[81] The signal intensity (current between electrode and enzyme) varies upon changing the potential, and a peak current is detected when the electrode potential matches the redox potential of the absorbed enzyme. In contrast to optical potentiometric titrations, cyclic voltammetry is not a pure thermodynamic technique and can also be used for the analysis of enzyme kinetics and mechanisms (see below; reviewed in ref. 80–82). The potential changes in PFV are not achieved by a chemical reductant and, thus, the technique is not restricted by the finite potential range provided by a chemical compound. Using a carbon electrode, the potential can be ramped between -1 V and $+800$ mV, while the lower potential limit for the reductant dithionite, for example, is only -550 mV at pH 7.0.[80] In comparison with the potentiometric solution techniques, an obvious disadvantage of PFV is posed by the experimental challenge of absorbing a protein monolayer onto the electrode surface without disturbing the native protein structure, changing conformational equilibria compared to the non-absorbed protein or affecting the enzymatic mechanism.[80]

3.1.3 Kinetics of PCET Reactions

Most approaches to describe PCET reactions mathematically have been developed using Marcus' semi-classical ET theory[83–89] as a starting point (see *e.g.* ref. 90–98). In the conventional form of semi-classical ET theory, long-range ET reactions are treated as non-adiabatic events and a classical concept of nuclear motion is employed, *i.e.* the nuclei of electron donor and acceptor are assumed to behave as harmonic oscillators, depicted as parabolas in graphic illustrations (Figure 3.4). In Marcus' original formalism (eqn (3.3)), the ET rate constant (k_{ET}) is proportional to an exponential term including the free activation energy ($\Delta G^{\ddagger}$) and to a pre-exponential factor k^0_{ET}, the rate constant for an activationless ET.

$$k_{ET} \propto k^0_{ET} \cdot \exp(-\Delta G^{\ddagger}/k_B T) \qquad (3.3)$$

The free activation energy is determined by the driving force $-\Delta G^{\circ}$, which is related to the standard redox potential difference (ΔE°), and by the so-called reorganisation energy λ (eqn (3.4))

$$\Delta G^{\ddagger} = \frac{(\Delta G^{\circ} + \lambda)^2}{4\lambda} \quad \text{and} \quad \Delta G^{\circ} = -n \cdot F \cdot \Delta E^{\circ} \qquad (3.4)$$

where n is the number of transferred electrons and F the Faraday constant. The activation parameter λ is the energy that is required to deform the nuclear geometries of the reactant atoms and their environment into the configuration of the product state without ET taking place.[99] It can be attributed to inner- and outer-sphere contributions. The inner-sphere reorganisation energy (λ_i) results from redox-dependent configurational changes in the redox centres, such as changes in bond lengths. Variances in the surrounding medium are reflected in the outer-sphere reorganisation energy (λ_o) and are, for instance, due to solvent reorientation or structural changes in the polypeptide backbone.[100]

The distance dependence of k_{ET} was subsequently introduced into the original classical Marcus theory by adding a quantum-mechanical electronic coupling element (H_{AB}). H_{AB} measures the strength of interaction between electron donor and acceptor at the nuclear configuration of the TS. Using a square-barrier tunnelling model as suggested by Hopfield,[101] H_{AB} decreases exponentially with the donor–acceptor distance (d, eqn (3.5))

$$H_{AB} = H^0_{AB} \cdot \exp\left[-\beta \cdot (d - d^0)\right] \qquad (3.5)$$

where H^0_{AB} is the overlap of the electronic wavefunctions at van der Waals distance ($d^0 = 3$ Å, ref. 88). The distance decay factor β, also known as the attenuation factor, depends on the conducting properties of the intervening medium between electron donor and acceptor (see *e.g.* ref. 84, 85, 89, 100,

102–105). Using this term for H_{AB}, k_{ET} can also be expressed in dependence of the donor–acceptor distance (eqn (3.6))

$$k_{ET} = k^{VdW}_{ET} \cdot \exp\left[-\beta \cdot (d - d^0)\right] \cdot \exp\left[-\frac{(\Delta G^\circ + \lambda)^2}{4\lambda k_B T}\right] \qquad (3.6)$$

where k^{VdW}_{ET} is the rate constant at van der Waals (VdW) distance (10^{13} s^{-1}) and corresponds to the characteristic frequency of the nuclei.[106] Including H_{AB} into eqn (3.3) finally yields the semi-classical ET equation, also known as the Marcus–Levich–Hush equation (eqn (3.7), ref. 85, 89, 101, 107, 108):

$$k_{ET} = \frac{2\pi}{h} \cdot H^2_{AB} \cdot FC = \frac{\pi}{\hbar} \cdot \frac{H^2_{AB}}{\sqrt{\pi \lambda k_B T}} \cdot \exp\left(-\frac{(\Delta G^\circ + \lambda)^2}{4\lambda k_B T}\right) \qquad (3.7)$$

where h is Planck's constant, $\hbar$ is Planck's constant divided by 2π, k_B is Boltzmann's constant and FC the Franck–Condon factor. If the nuclei are treated as harmonic oscillators, FC describes the nuclear overlap of reactant and product functions. Maximum rate constants are obtained for $-\Delta G^\circ = \lambda$.

The difficulty in studying biological ET systems consists of ascertaining that the observed rate constant (k_{obs}) reflects a true ET event (k_{ET}).[109] Biological redox reactions can involve a multitude of steps leading up to, imbedding or accompanying the actual ET. The redox mechanism may require protein–protein interactions, conformational changes or chemical reactions such as de-/protonation and hydrogen transfer steps. These adiabatic reaction steps may be essential in producing an activated protein complex capable of kinetically and/or thermodynamically efficient ET (see * in eqn (3.8)).

$$
\begin{aligned}
&A_{ox} + B_{red} \overset{K_a}{\rightleftharpoons} A_{ox}/B_{red} \overset{k_x}{\underset{k_{-x}}{\rightleftharpoons}} [A_{ox}/B_{red}]^* \overset{k_{ET}}{\underset{k_{-ET}}{\rightleftharpoons}} A_{red}/B_{ox} \\
&A_{ox} - B_{red} \overset{k_x}{\underset{k_{-x}}{\rightleftharpoons}} [A_{ox} - B_{red}]^* \overset{k_{ET}}{\underset{k_{-ET}}{\rightleftharpoons}} A_{red} - B_{ox}
\end{aligned}
\qquad (3.8)
$$

Models accounting for the kinetic complexity of biological ET distinguish between true, gated and coupled ET (eqn (3.9), (3.10) and (3.11); ref. 100, 106, 110, 111).

$$\text{true ET:} \quad k_{ET} \ll k_x \quad K_x = k_x/k_{-x} \gg 1 \quad k_{obs} = k_{ET} \qquad (3.9)$$
$$\text{gated ET:} \quad k_{ET} \gg k_x \qquad\qquad\qquad\qquad k_{obs} = k_x \qquad (3.10)$$
$$\text{coupled ET:} \quad k_{ET} \ll k_x \quad K_x = k_x/k_{-x} \ll 1 \quad k_{obs} = K_x \cdot k_{ET} \qquad (3.11)$$

In a true ET reaction (eqn (3.9)) the ET will be the slowest and overall rate-limiting step. Experimentally observed λ-values report solely on ET and are

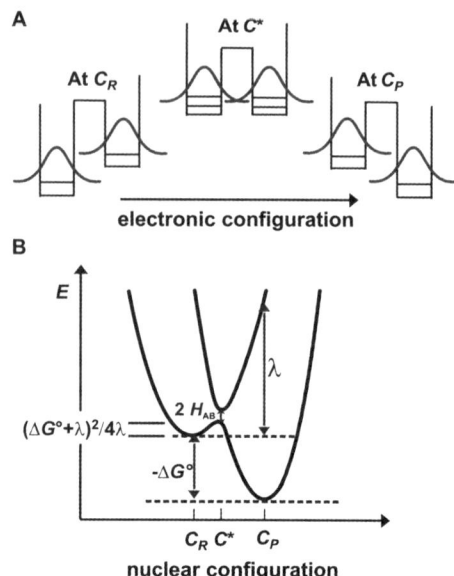

Figure 3.4 Electronic energy profiles for an ET reaction treated quantum-mechanically (A) and the corresponding nuclear coordinate (B). The three panels in (A) show the potential energy surface as a function of the electronic coordinate at three values of the nuclear coordinate (B): C_R is for the reactant state, C^* for the transition state (TS) and C_P for the product state, respectively. The horizontal lines in panel A represent the ground state vibrational wavefunction of the electron. At the configuration of the TS, the reactant and product ground states are transiently degenerate resulting in an overlap of the electronic wavefunctions, which is proportional to the electronic coupling term H_{AB}. For a non-adiabatic ET reaction, H_{AB} is weak (small). $\Delta G°$ is the driving force, λ the reorganisation energy. Adapted from ref. 241.

expected to be in the region of 1 eV. For a true ET, the electronic coupling between donor and acceptor is weak and the non-adiabatic limit has been suggested to be 80 cm^{-1}.[106,112] In a gated ET step[109,113] (eqn (3.10)), a reaction preceding the ET rate-limits the ET process and the observed rate constant (k_{obs}) only reflects the non-ET reaction. The analysis of k_{obs} will not depend on the driving-force, since the observed step is not fueled by the redox potential gradient. Unreasonably large values of H_{AB} (*e.g.* 10^6 cm^{-1}; ref. 114) and λ (*e.g.* 4 eV; ref. 114), negative donor–acceptor distances d (*e.g.* ref. 114) and driving-force independent λ-values[109] are indicative of a gated ET reaction. A coupled ET reaction (eqn (3.11)) is characterised by a fast, but unfavourable pre-equilibrium and the k_{obs} value is predicted to be a product of the equilibrium constant (K_x) of the non-ET step and the true ET rate constant (k_{ET}). Such a coupled ET may still exhibit a predictable dependence on $\Delta G°$; reasonable values for H_{AB} and λ may be obtained, which would also be consistent with a true ET.[115] In the context of this review, examples for a kinetically gated or coupled ET reaction are an ET preceded by a PT step, that either rate-limits the ET step (gated ET) or poses an unfavourable pre-equilibrium (coupled ET) thereby decreasing the k_{obs} value. For the kinetic mechanism being adequately described by eqn (3.9)–(3.11), the ET and PT steps have to be separate elementary reactions with two corresponding TS. In other words, such kinetically gated and coupled ET reactions are included in the stepwise mechanisms of PCET reactions (see Figure 3.1).

Concerted PCET reactions cannot be described solely by the semi-classical Marcus ET theory but necessitate special kinetic treatments, as both protons and electrons affect the coupling matrix H_{AB} and the Franck–Condon (*FC*) factor (eqn (3.7)). Due to the wave-particle duality, protons and electrons can tunnel – albeit with different characteristics, since the wave-like properties are more pronounced for lighter particles. Apart from the opposite charge of electrons and protons, the difference in mass is one of the main distinctive features, with the proton being ~ 2000 times heavier than an electron.[20,25,84,89,112] The de Broglie wavelength, which is a measure of the distance a particle can tunnel, is inversely proportional to the square root of the particle's mass and is ~ 40 times smaller for a proton (deuteron) than for an electron.[50] Typically, a proton can tunnel over distances in the region of a hydrogen bond,[116,117] whereas an electron can overcome ~ 17 Å on a µs time scale, if the reaction is driving-force optimised (see above, $-\Delta G^{\circ} = \lambda$; ref. 85). As a result, even small bond vibrations reducing the proton donor–acceptor distance can play a decisive role (see below) and a distinct orbital pre-orientation is paramount in PT reactions. The different requirements and abilities of protons and electrons are especially well accommodated in bi-directional PCET mechanisms. In HAT reactions, in contrast, the electron stays "associated" with the proton and the maximum transfer distance is governed by the heavier particle. In such cases, the electronic coupling element H_{AB} can be large and the ET may be electronically adiabatic, *i.e.* the transferred electron is always in equilibrium with the nuclear motions, creating the ET barrier.[94]

It is worth noting that a PCET does not require the breaking/making of an X–H covalent bond. According to the thermodynamic redox-Bohr effect (see Section 3.1.2), any relative movement of the electron and the proton affects their mutual influence, *i.e.* any electronic motion results in an altered acidity and any proton fluctuation, such as a bond vibration, yields a changed redox potential.[9] This sensitive interference does not only have thermodynamic but also kinetic consequences, which can lead to an observed PCET event, even if only one charge is actually transferred during the process.[20] One example is the long-range ET through a network of hydrogen bonds, such as the ET pathways found in proton pumps like cytochrome *c* oxidase.[20] In these cases, the coupling emerges from O–H bond vibrations influencing the ET rate constant. The sensitivity of PT towards the transfer distance is also responsible for the high substrate and reaction specificity of enzyme-catalysed transformations.[21]

The different characteristics of PT and ET events and their mutual influence discussed in the previous paragraph have to be accounted for in any mathematical approach. It is beyond the scope of this review to introduce the reader to the various theoretical models that have been developed on the basis of the semi-classical ET theory and of related PT formalisms (*e.g.* ref. 90–98; reviewed *e.g.* in ref. 118). Certain features, that are crucial for the understanding of experimental observations, will be outlined in the following. One complicating aspect is the requirement for different mathematical assumptions when analysing a PCET reaction, *e.g.* different equations have to be derived for HAT, HIT and non-HAT/non-HIT PCET. While HAT and HIT imply small ET distances and, hence, an electronically adiabatic ET, PCET with long-range ET are typically electronically non-adiabatic (see Marcus theory above).[94,97,119] In principle, however, PCET mechanisms can be electronically adiabatic, non-adiabatic or a mixture of both depending on the relative rate constants of PT and ET.[97]

Hammes-Schiffer and co-workers have derived a PCET equation for an electronically adiabatic PT and an electronically non-adiabatic ET.[120–123] In this approach, the distance sensitivity of PT is accounted for by summing over the reactant and product vibronic states and weighting them by the Boltzmann factor for the thermal population of each vibrational mode. The accessibility of a certain channel will therefore depend on the thermally available energy, *i.e.* on the temperature, and on the energy gap between adjacent vibrational quantum levels ($\Delta E = h\nu$, with the frequency ν and Planck's constant h). Due to the variations in the PT distance, each vibrational channel is characterised by different electronic coupling constants, reorganisation energies and free energy changes.[93,94,96] The provided framework allows the analysis of experimental data *a posteriori*, while the theory cannot be employed to predict PCET rate constants.[21,93]

Klinman and co-workers refined a model originally developed by Kuznetsov and Ulstrup,[124] which considers fluctuating H-transfer distances on the time scale of the PCET.[117,125,126] This is in contrast to the concept by Hammes-Schiffer discussed above, in which the overlap of the proton vibrational wave function is constant for the time period of the PCET. Klinman's concept was brought forward after experimental findings on the kinetic isotope effects (KIE *e.g.* $= k_H/k_D$) of several enzymatic systems could not be reconciled with established tunnelling corrections expanding the semi-classical transition state theory (Figure 3.5 *e.g.* ref. 116,117,127–134; see Section 3.3.2). The observations were explicable, however, when assuming a considerable contribution of promoting vibrations/motions that modulate the tunnelling barrier. The rate constant of these environmentally coupled non-adiabatic H-transfer reactions is predicted to depend on three components (Figure 3.5; ref. 117,125,126): in addition to an electronic coupling term and a Marcus-like expression describing the barrier, a so-called Franck–Condon (*FC*) gating term was introduced, which takes the fluctuating barrier into consideration (distance sampling). The *FC* term depends on the mass, the frequency and the transfer distance of the proton as well as on the properties of the fluctuating barrier. The vibrating barrier is approximated by a harmonic oscillator with a gating energy E_X. If the thermally available energy ($k_B T$) approaches E_X, the promoting vibrations can modulate the H-transfer distance resulting in faster rate constants and a small, temperature dependent KIE.[118] In the case where the promoting motions are

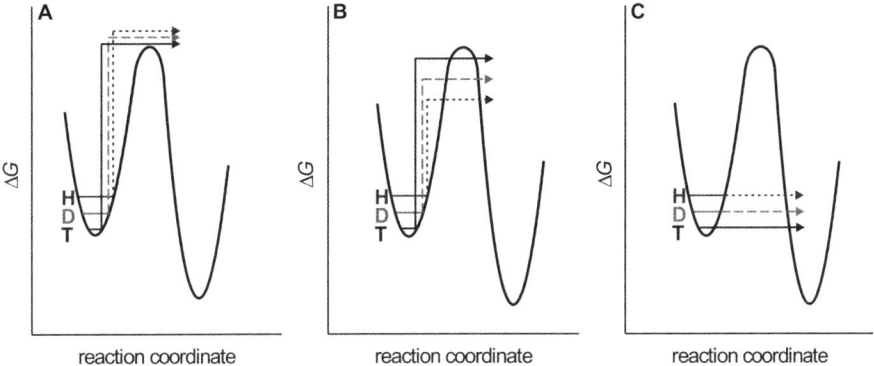

Figure 3.5   Models accounting for kinetic isotope effects (KIE). (A) Principle of the KIE according to the semi-classical transition state theory (TST; ref. 242). The vibrational ground states within the potential energy well of the reactant state are presented as horizontal bars for hydrogen (H), deuterium (D) and tritium (T) indicating the differences in their zero-point energies ($ZPE = 1/2h\nu$). According to the TST, the KIE is due to the different ZPE and the semi-classical limit for the KIE is ~ 7. In the semi-classical regime, the Arrhenius pre-factors (A_H/A_D) typically yield a ratio of ~ 1 and the difference in the activation energies ($\Delta E_a = E^D_a - E^H_a$) can be large. The KIEs are modestly temperature dependent. (B). The Bell tunnelling model.[243] The Bell correction reflects the ability of the reactants to form products *via* tunnelling, which is larger for lighter particles (H > D > T). Tunnelling occurs just below the TS and heavier isotopes may form products by traversing more closely to the TS and/or by over-the-barrier transfer. Indicators of H-tunnelling are a non-linear Arrhenius plot, KIE values of $k_H/k_D > 7$, small Arrhenius-prefactors ($A_H/A_D < 1$), ΔE_a values of up to 5–6 kJ mol^{-1} and highly temperature dependent KIEs. (C). Deep tunnelling. In the regime of deep tunnelling, H-transfer occurs far below the reaction barrier[118] and is well described by full quantum-mechanical tunnelling models.[124,156,230,244] The rate constants are temperature independent and moderately temperature dependent KIEs with $KIE = A_H/A_D$ are obtained. Please note, that none of these schemes presented in panels A–C accounts for environmentally coupled H-transfer reactions. Adapted from 117 and 118.

energetically inaccessible ($E_X \gg k_B T$), temperature independent KIEs can be observed. The gating term in Klinman's formalism thus accounts for the coupling between environmental dynamics and the reaction coordinate of PCET reactions.

3.2   Experimental Kinetic Approaches to Analyse PCET Reactions

In this review, the phrase PCET is used to comprise kinetically gated and coupled stepwise mechanisms as well as concerted PCET reactions. The reasons for this broad definition are rooted in the experimental challenge of

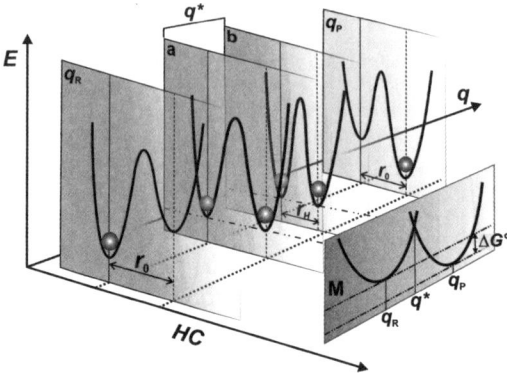

Figure 3.6 Promoting vibrations in an environmentally coupled hydrogen transfer reaction. The axes are: E, energy; HC, hydrogen coordinate; q, environmental coordinate. The four large panels represent the potential energy curve as a function of the hydrogen coordinate at three different values of the environmental coordinate: q_R, reactant state; q^*, transition state; q_P product state. The ground-state vibrational wavefunction of the H-nucleus is illustrated by the grey spheres. The single panel labelled M gives a Marcus-like perspective of the free energy curves in dependence of q. $-\Delta G°$ is the driving force of the reaction. The symmetry of the double-well is modulated by environmental motions, resulting in a system with almost degenerate quantum states at the transition state ($q = q^*$). In this configuration, the hydrogen can tunnel between the wells. Due to a gating motion along the hydrogen coordinate, the tunnelling distance in panel b is reduced to r_H compared to the larger equilibrium value r_0 in panel a. Adapted from 132 and 245.

distinguishing between stepwise and concerted events. To follow an enzymatic ET reaction kinetically, the colour changes accompanying the reduction/oxidation of certain redox-active cofactors are frequently used as a probe. Prominent examples are heme groups, flavins and metal sites that are composed of, for example, transition state metals such as Cu ions coordinated by amino acid residues. One option to prove the formation of a chemical intermediate in a PCET reaction is the detection of an enzyme species, which has either a changed redox state or an altered protonation state and which is spectroscopically distinct from the enzyme species resulting from both events. This requirement of observing PT and ET steps separately is rarely fulfilled. Another complication can arise even if a probe for an intermediate is available: the chemical intermediate may not sufficiently accumulate kinetically during the enzymatic reaction. This scenario is in fact inherent in the stepwise PCET mechanisms. If an ET reaction is *e.g.* rate-limited (gated) by a preceding slow PT step, an eventual intermediate will be difficult to detect due to its slow build-up and the subsequent rapid depletion. In the case of a stepwise kinetically coupled PCET, a fast, but unfavourable, pre-equilibrium may result in a very low concentration of the intermediate species thus giving a poor signal. In the following, several kinetic tools will be introduced, which (i) can be used to test for a PCET in general and (ii) may occasionally reveal kinetic evidence that allows for the discrimination between gated, kinetically coupled and concerted mechanisms.

Solvent kinetic isotope effects (SKIE) and KIEs belong to the most frequently employed techniques to test for a PCET. From a kineticist's viewpoint, PCET reactions could actually be defined as "an ET, that exhibits a (S)KIE", as this experimental evidence automatically includes both stepwise and concerted PCET mechanisms. The observation of a SKIE reports on the contribution of solvent molecules, buffer components or solvent-exchangeable protons to the rate-limiting step(s).[135–137] Further information can be gathered by varying the solvent from 0–100% D_2O. In these so-called proton inventory experiments,[137] the dependence of the observed rate constant on the fractional solvent composition can provide evidence for the number of protons/solvent molecules contributing to the rate-limiting step. A linear proton inventory plot is indicative of a single proton/solvent molecule, while an observed curvature suggests the involvement of more than one proton/solvent molecule.[137] For KIE studies, specific atoms are isotopically substituted, *e.g.* the abstracted hydrogen of a substrate C–H bond (*e.g.* ref. 138, 139) or the hydrogen atom in the NAD(P)H cofactor (*e.g.* ref. 72, 140–142). By stereo-specifically labelling the *pro*-(*R*) or *pro*-(*S*) hydrogen in NAD(P)H, the primary and/or secondary KIE of the hydride transfer reaction can be determined yielding further mechanistic insights (*e.g.* ref. 72, 141, 142). The temperature dependence of (S)KIEs can be a useful diagnostic tool to determine the tunnelling regime for the reaction and to test for promoting motions (see above, Figures 3.5 and 3.6, and Section 3.3.2).

(S)KIE studies can be performed in the steady-state (*e.g.* ref. 134, 138, 139, 143–165) as well as using rapid kinetic techniques, such as the stopped-flow and the laser instrument (*e.g.* ref., 72, 118, 127, 132, 140, 165–183). Single-turnover experiments are advantageous in that the reductive half-reaction of a ping-pong redox mechanism can be analysed separately. In general, the observed exponential traces can be easily fitted, yielding single or multiple observed rate constants, which can potentially be assigned to distinct reaction steps. The mechanistic implications of the data can, however, be challenging to ascertain, especially when a complex kinetic mechanism results in a limited number of exponential phases due to kinetic mixing of several reaction steps or of parallel pathways (see case study on nitrite reductase below). Steady-state studies give catalytic rate constants (k_{cat}), the Michaelis–Menten constants (K_m) and the specificity constant (k_{cat}/K_m), a second-order rate constant. The KIE on the k_{cat} value measures the contribution of a PT step (PT steps) to the rate-limiting catalytic event and is of finite diagnostic value. A multiple-turnover KIE experiment conducted with the isotopes hydrogen (H), deuterium (D) and tritium (T) does allow the numeric calculation of intrinsic KIE values (*e.g.* k_H/k_D) for a distinct reaction step from the observed competitive specificity constants[184] $^T(k_{cat}/K_m)_{H\,obs}$ and $^T(k_{cat}/K_m)_{D\,obs}$.[135,184] However, the involved mathematical procedure is rather more complex than the calculation of the KIE values in single-turnover experiments. If the steady-state experiments are performed using all three isotopes, the Swain–Schaad exponents can provide further mechanistic insight.[185,186]

The interpretation of SKIEs can be difficult, since the macroscopic effect may result from a combination of solvent/buffer molecules and/or amino acid side chains acting as acid–base catalysts. Conducting a pH dependence of the observed kinetic process may provide useful indications of a potential acid–base catalyst and kinetic pK_a values can be compared with thermodynamic pK_a values determined, *e.g.* in redox titrations (see above). With the increasing availability of crystal structures, active site residues of potentially catalytic significance are more readily identifiable thus allowing systematic mutational studies. Even though active site mutations are of great diagnostic value, complications can arise when solvent/buffer components or neighbouring amino acid residues are able to compensate for the lost catalytic function.[29] In some cases, a macroscopic pK_a may not be attributable to a single residue at all, as a group of closely spaced proton donors and acceptors may catalyse the reaction cooperatively.[29] The involvement of buffer components as acid–base catalysts yields pH-dependent rate constants and can be further investigated by varying the buffer concentration.[29,187] Another complicating feature to be aware of has been pointed out for reactions catalysed by solvent molecules (*e.g.* H_2O, OH^-, H_3O^+),[29,187] as the observed rate constant may not necessarily exhibit a pH dependence. A pH-dependent forward rate constant is expected, when OH^- (H_3O^+) acts as base (acid). This effect does not result from a pH-dependent driving force, but is due to OH^- (H_3O^+) participating as a substrate.[29,187] In contrast, no pH-dependent behaviour is observed for the forward reaction, if H_2O functions as acid–base catalyst. The backward reaction, however, does depend on the pH value, since H_3O^+/OH^- is involved as a reactant.[29,187]

To tackle the above discussed challenge in detecting potential PCET intermediates, an intricate experimental approach has been developed to measure the rate constant of the PT step independently from the ET (see *e.g.* ref. 188–193). When performing experiments in unbuffered solutions, pH indicators, such as phenol red, can be used as reporters of the PT step. The rate constants obtained from the colour change of the dye correspond to the PT of interest, assuming that the equilibration between the protons in the active site and the bulk solvent is not rate-limiting.

The mechanism of hydride transfer reactions is of special biochemical interest due to the enzymatic significance of the cofactor NAD(P)H (see 3.2). In case of a stepwise PCET reaction, short-lived radical intermediates are expected[44–46] and the existence of unpaired electrons may be probed by analysing the magnetic-field effect on the kinetics using novel stopped-flow technology.[194–196] In a modified Applied Photophysics SX.18MV-R reaction analyser, an external magnetic field of up to 75 mT can be applied.[194–196] Since potentially formed radical intermediates in a stepwise hydride transfer mechanism are expected to be short-lived,[44] this approach is especially useful, as the magnetic field effect is an indirect probe of the radical formation. In future, an analogously modified laser photoexcitation instrument might expand the scope of this novel approach by improving the time resolution by three orders of magnitude to the µs time scale. Another kinetic technique to measure the transient formation of radical species is time-resolved electron paramagnetic resonance (EPR). Although not widely applied for the analysis of ET reactions, several transient EPR studies have been published.[197–201]

Protein film voltammetry (PFV) at low scan rates has been introduced above as an equilibrium technique to determine redox potentials of adsorbed enzymes. In the presence of substrate, however, catalytic turnover can take place, producing sigmoidally shaped wave functions in a current *versus* electrode potential plot.[81,82] The sigmoidal behaviour reflects the increase of electron flux between the electrode and the protein, as the potential is raised. A maximal current (*i.e.* rate) is observed when the potential provides a sufficient driving force for the enzyme to remain in a steady-state. By modulating the substrate concentration, the Michaelis–Menten parameters K_m and k_{cat} can be determined and further insight is obtained by varying the pH value or the temperature. In addition to providing equilibrium and steady-state data, PFV can be employed as a rapid kinetic technique to detect transient catalytic intermediates.[81,82] In these fast voltammetric experiments, the electrode potential is rapidly modulated (scan rates of up to 1000 V s^{-1},[202] rather than mV s^{-1}, in the steady-state) transforming the sigmoidal waveform observed in the steady-state into a peak-like signal. At sufficiently high potential modulations, electrons flow into the redox centres of the enzyme, but product formation does not take place, since the electrode "reclaims" the provided electrons before they are delivered to the substrate. In addition to standard rate constants for electron exchange resulting from these experiments, kinetically limiting reactions coupled to the ET step, such as a PT, can also be scrutinised. By performing non-turnover voltammetric experiments at various scan rates, the peak positions are plotted *versus* the log(scan rate) giving a trumpet-like shape.[29,202] In the case of a PCET, a distorted plot is obtained, which can then be modelled in relation to the rate constant and the thermodynamics of the additional reaction.[22] Using scan rates of up to 1000 V s^{-1} allows the detection of coupling events on the sub-ms time scale.[202] In cytochrome *c* nitrite reductase, a PCET mechanism was established using a combination of catalytic and non-catalytic PFV.[203]

3.3   Case Studies

3.3.1   PCET in Nitrite Reductase

Dissimilatory nitrite reductases (NiR) play a pivotal role in the anaerobic respiration cascade of denitrifying bacteria, archaea and fungi by catalysing the first committed step of the pathway.[204–206] The net reaction of NiR yields the conversion of nitrite (NO_2^-) into gaseous nitric oxide (NO) and water (H_2O) (eqn (3.12); ref. 207):

$$NO_2^- + 2H^+ \rightleftharpoons NO + H_2O \qquad (3.12)$$

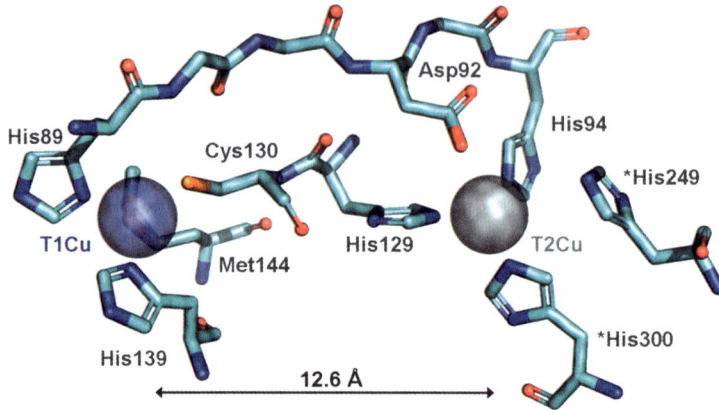

Figure 3.7 Structure of the T1Cu-T2Cu site in *Ax*NiR, a blue CuNiR (PDB 1oe1). His89, Met144, Cys130 and His139 ligate the T1Cu centre. Met144 is presented in two conformations. The T2Cu is coordinated by His129, His94 and *His300. The two Cu centres, shown as spheres, are linked by Cys130 and His129. Asp92 and *His249 have been suggested to act as acid–base catalysts. The stars (*) denote residues provided by the adjacent enzyme subunit. For clarity, only the side chains of key amino acid residues are shown.

Copper-containing NiRs (CuNiR) utilise two copper centres to catalyse the one-electron reduction of nitrite. In NiR from *Alcaligenes xylosoxidans* subsp. *xylosoxidans* (NCIMB 11015, *Ax*NiR), the non-catalytic Cu site (T1Cu) is blue in the oxidised state and becomes colourless upon reduction by the natural electron donor protein azurin.[208] Subsequently, the electron is transferred from the T1Cu to the colourless catalytic Cu centre (T2Cu) *via* a 12.6 Å covalent bridge thereby restoring the blue oxidised T1Cu (Figure 3.7, ref. 209–211). According to eqn (3.12), two substrate protons are consumed during the redox process, given that H_2O rather than OH^- is formed as a product. Several catalytic mechanisms have been proposed (see *e.g.* ref. 212, 213) and the amino acid residues Asp92 and His249 have been suggested to function as acid–base catalysts yielding maximum catalytic turnover between pH 5.2 and 6.[211,214–218]

The T1Cu to T2Cu ET in NiR has been extensively analysed by pulse radiolysis using the colour change of the blue T1Cu ($\varepsilon_{600} \sim 6.3\ \mathrm{mM}^{-1}\ \mathrm{cm}^{-1}$ per NiR trimer) with observed ET rate constants (k_{ETobs}) ranging from $\sim 150\ \mathrm{s}^{-1}$ to $\sim 2000\ \mathrm{s}^{-1}$.[207,208,215,216,219–221] Due to the absence of any systematic analysis of a potential PCET mechanism in NiR, we examined the coupling of ET and PT steps in *Ax*NiR both in the steady-state and in single-turnover reactions at pH 7.0.[222] In a first set of experiments, the net consumption of protons was determined in unbuffered solution using the pH indicator phenol red. Two protons were confirmed to be turned over per nitrite molecule, indicating the formation of H_2O rather than OH^-. The catalytic turnover of *Ax*NiR measured in H_2O and D_2O yielded classic substrate concentration dependencies with comparable Michaelis–Menten constants around 35 μM (Figure 3.8A). A moderately elevated SKIE of 1.3 ± 0.1 was observed, indicative of one proton or one solvent molecule contributing to the rate-limiting catalytic step(s) (Figure 3.8B). In contrast, a complex behaviour was revealed, when the substrate concentration dependence of the inter-Cu ET rate constant (k_{ETobs}) was analysed in laser photoexcitation experiments (Figure 3.8A). While inter-Cu ET in D_2O gave a "normal" hyperbolic curve with accelerated rate constants in the presence of substrate, the observed inter-Cu ET rate constants decreased in H_2O with increasing nitrite concentration up to ~ 10 mM. At substrate concentrations > 10 mM, the rate constants approached the value detected in substrate-free AxNiR. Since the rates observed in D_2O exceeded those in H_2O above 0.1 mM nitrite, a substrate-concentration dependent SKIE was obtained with inverted values above this concentration (Figure 3.8B). Proton inventory experiments performed in the laser instrument at 10 mM nitrite substantiated the inverse SKIE (Figure 3.8C) and the linear behaviour suggests the involvement of one proton/one solvent molecule in the observed ET reaction. For the substrate-free inter-Cu ET, a normal SKIE could be confirmed with one proton/one solvent molecule rate-limiting the ET step, *i.e.* inter-Cu ET is accompanied by a PT even in the absence of product formation. A comparison of the catalytic rate constants with the k_{ETobs}-values shows that the PT affecting the catalytic turnover cannot be equivalent with the PT step responsible for the unusual SKIE detected in the single-turnover experiments.

The underlying kinetic mechanism is undoubtedly complex. One explanation might be provided by a random-sequential reaction sequence analogous to that suggested in ref. 223, in which the substrate preferentially binds to the enzyme species with a reduced T1Cu centre, but is also capable of ligating to the fully oxidised AxNiR (Figure 3.9). For the laser experiments, this scenario would imply a pre-equilibrium between nitrite-bound and substrate-free AxNiR species prior to the reduction upon photoexcitation. Hence, parallel pathways might exist for which the relative kinetic and/or thermodynamic population is potentially a function of both the solvent and the substrate concentration. This mechanistic complexity may ultimately result in the unusual trend observed for the SKIE. The temperature dependencies of k_{ETobs} in H_2O and D_2O in the presence and absence of 10 mM nitrite did not yield Marcus parameters that were in disagreement with a true ET, thus making a gated ET unlikely. Since the SKIE clearly proved an ET reaction coupled to PT, the complex kinetic behaviour in AxNiR is most likely a result of a kinetically coupled stepwise PCET or a concerted PCET.

The presented case study on inter-Cu ET in AxNiR demonstrates the experimental challenges associated with PCET reactions. The existence of multiple pathways renders any mathematical analysis rather complex, especially if only one macroscopic parameter – as in this case k_{ETobs} – is experimentally accessible. All attempts to employ the pH indicator phenol red in the laser experiments to obtain rate constants of the PT steps (k_{PTobs}) failed, which we attribute to potential photo-reduction and/or photo-degradation processes of the dye. Due to

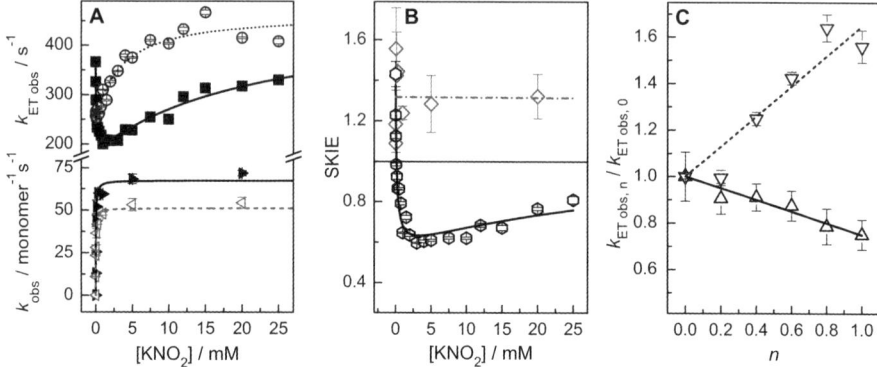

Figure 3.8 Kinetic isotope effect studies of inter-Cu ET and catalytic turnover in *Ax*NiR. Single-turnover data were obtained in laser photoexcitation experiments using NADH as artificial electron donor and monitoring the blue Cu absorbance signal at 600 nm. The multiple-turnover studies were conducted using a novel assay with hemoglobin as a NO-sensor. (A) Substrate concentration dependence of the catalytic turnover (k_{obs}/monomer *Ax*NiR; triangles) and the observed inter-Cu ET rate constant (k_{ETobs}, squares and circles) in phosphate-buffered H_2O (pH 7.0, 4 °C, closed symbols) and D_2O (pD 7.0, 4 °C, open symbols). The catalytic turnover data were fitted to a classical Michaelis–Menten equation, the D_2O laser data set was fitted using an offset hyperbolic equation and the H_2O laser data were analysed by an equation adapted from steady-state inhibition schemes. (B) Observed SKIE (k_{H2O}/k_{D2O}) for the single- and multiple-turnover of *Ax*NiR (black pentagons and grey diamonds, respectively). The catalytic SKIE was fitted linearly (dash-dotted line), while the transient SKIE was analysed using an equation adapted from steady-state inhibition schemes. (C) Proton inventory laser experiments in the absence of substrate (grey down-triangles) and in the presence of 10 mM nitrite (black up-triangles). The ratio of the observed rate constant $k_{2obs,n}$, obtained at a certain volume fraction of D_2O (n), and the rate constant $k_{2obs,0}$ in pure H_2O is plotted *versus n*. The lines are linear fits. See ref. 222 for details.

these experimental limitations, the distinction between the two PCET mechanisms – a kinetically coupled stepwise or a concerted PCET – remained elusive. Mutation analyses, including *e.g.* the suggested acid–base catalysts Asp92 and His249, may result in altered rate constants, but it is likely that other amino acid residues in the active site and/or water molecules will (partially) compensate for the mutations. Therefore, any resulting effects have to be interpreted with caution and are unlikely to reflect an on-off switch of the PCET mechanism.

3.3.2 Hydride Transfer Reactions in Old Yellow Enzymes

The Old Yellow Enzyme (OYE) family of flavoproteins consist of a large group of flavin mononucleotide (FMN)-containing NAD(P)H-dependent oxidases. We will describe our work on two of these enzymes: morphinone reductase

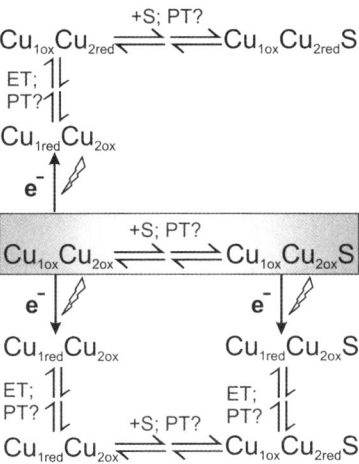

Figure 3.9 Random mechanism suggested for NiR. Cu_{1ox}, oxidised blue T1Cu; Cu_{1red}, reduced colourless T1Cu; Cu_{2ox}, oxidised T2Cu; Cu_{2red}, reduced T2Cu: ET, electron transfer; PT, proton transfer; S, substrate; e^-, electron. The grey box symbolises the pre-equilibrium likely to exist prior to photoexcitation (flash) in the laser experiments. Please note that the order of the reaction steps labelled "$x;y$?" is unknown and that all reaction steps may be concerted rather than sequential as suggested by the arrows. The ET reaction may also be coupled to subsequent PT steps.

(MR) isolated from *Pseudomonas putida* M10 and pentaerythritol tetranitrate reductase (PETNR) from *Enterobacter cloacae*. Catalysis by MR and PETNR proceeds in two sequential half-reactions, denoted the reductive (RHR) and oxidative (OHR) half-reactions[142] (Figure 3.10). By monitoring the oxidation state of the FMN, the RHR and OHR can be individually probed experimentally using rapid-mixing techniques such as stopped-flow spectrometry.[142,224] The RHR in MR and PETNR is identical and involves stereoselective hydride transfer from the C4 pro-*R* hydrogen of NADH (MR) or NADPH (PETNR) nicotinamide moiety to the FMN-N5 atom (Figure 3.10). The H-transfer appears to be concomitant with electron transfer to the FMN as there are significant isotope effects of ~ 5 observed[142,225] – *i.e.* H-transfer is at least partly rate-limiting during FMN reduction, hence this is an apparent PCET reaction. In the OHR, MR and PETNR are able to reduce a number of substrates including unsaturated compounds and nitroamines.[224,226] The OHR involves formal hydride transfer from the N5 atom of the reduced FMN to the oxidising substrate. In reactions where carbon double bonds are reduced, the OHR also involves proton transfer from solvent to the oxidising substrate.[140] Isotope effects are observed upon reoxidation when the FMN-N5 proton is deuterated (KIE ~ 3.5, ref. 140) and when the reaction with unsaturated substrates is performed in deuterated solvent (KIE ~ 2.3, ref. 140). Again, this suggests that both the hydride and proton transfers are at least partially rate-limiting during FMN oxidation.[140]

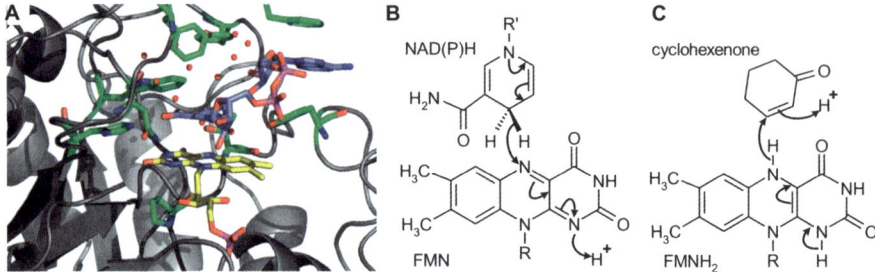

Figure 3.10 The reductive half-reaction of MR. (A) The X-ray crystal structure (pdb 2R14, ref. 141) of the active site of NADH$_4$-bound MR showing the FMN in yellow, NADH$_4$ in purple and key active site residues in green. Water molecules within 6 Å of the NADH$_4$-C4 are shown as red spheres. (B, C) The reaction mechanisms of the RHR of MR/PETNR with NAD(P)H (B) and the OHR with cyclohexenone, a generic unsaturated substrate (C). Note: the proton abstraction from solvent by the reduced FMN-N1 during the RHR (B) is not rate-limiting as there is no observed SKIE.[180] However, the proton abstraction by the activated cyclo-hexenone during the OHR in MR appears to be partially rate-limiting, as there is an observed SKIE of 2.3.[140] The hydride transfer in both half-reactions is also at least partially rate-limiting, as KIEs are also observed.[142,228,229]

We have recently used QM/MM computational methods to show that the H-transfer during the RHR of MR with NADH occurs predominantly ($>99\%$) by quantum mechanical tunnelling.[227] In this case, the H-transfer does not occur *via* a classical transition state (over-the-barrier) but rather tunnels about 25 kJ mol^{-1} below the top of the reaction barrier (which is ~ 80 kJ mol^{-1} high).[227] This is considered a 'deep' tunnelling reaction (see Figure 3.10). Experiments have shown that the KIE on this reaction is strongly temperature dependent ($\Delta\Delta H^{\ddagger} \sim 8$ kJ mol^{-1})[142,228,229] and we have interpreted these data such that there is a thermally-activated fast (sub-ps) promoting vibration (Figure 3.10, ref. 124, 156, 230 and 231) that dynamically compresses the H-transfer reaction barrier width, thus enhancing the rate of H-transfer.[132,232] Within the active site of the binary complex of MR/PETNR and NAD(P)H, the FMN isoalloxazine and NADH nicotinamide rings are roughly coplanar forming a charge-transfer (CT) complex (Figure 3.10, ref. 141 and 224). We have shown that high pressure appears to effectively 'squeeze' the isoalloxazine and nicotinamide rings together within the MR active site.[233] This 'squeezing' reduces the average NADH-C4 (heavy atom donor) and FMN-N5 (acceptor) separation, and appears to cause the significant H-transfer rate enhancement.[181,233] Together, these data show that the rate of the hydride transfer during the RHR of MR with NADH appears to be fully rate-limited by the H-transfer. As the RHR is expected to be quite exothermic[180] and the edge-to-edge electron transfer distance will be <5 Å (Figure 3.10), the rate of pure electron transfer should be extremely fast. Consequently, this is not surprising.

3.4 Concluding Remarks

In this chapter, we have defined PCET reactions as an ET step with an experimentally observed (S)KIE. This broad definition comprises stepwise (kinetically gated or coupled) PCET reactions as well as concerted PCET mechanisms (see Figure 3.1) and is a consequence of the experimental challenge to distinguish between stepwise and concerted mechanisms. In addition to the technical constraints currently limiting the detailed kinetic analysis of PCET reactions, intrinsically complex enzymatic mechanisms may be the reason for contentious experimental findings and may complicate any mechanistic conclusions. Some examples for mechanistically complex systems include: (i) the kinetic mixing of successive steps resulting in a smaller number of observed pre-steady-state rate constants than reaction steps; (ii) the (kinetic) mixing of parallel reaction pathways (see Section 3.3.1); (iii) conformational subsets of the enzyme-substrate complex resulting in a multi-dimensional free energy landscape with a multitude of local energy minima and transition states and giving a larger number of observed rate constants than reaction steps expected from the catalytic mechanism.[234,235] In the future, the development of novel and advanced kinetic approaches may facilitate the independent determination of H-transfer and electron transfer steps. The desirable experimental progress may also provide further insights into net hydride ion transfer reactions and may answer the unresolved issue of whether enzymatic hydride ion transfer (HIT) does compete with other PCET mechanisms such as ET–PT–ET or HAT–ET.

References

1. P. Mitchell, *Nature*, 1961, **191**, 144.
2. R. Neutze, E. Pebay-Peyroula, K. Edman, A. Royant, J. Navarro and E. M. Landau, *Biochim. Biophys. Acta*, 2002, **1565**, 144.
3. B. E. Ramirez, B. G. Malmstrom, J. R. Winkler and H. B. Gray, *Proc. Natl. Acad. Sci. U. S. A.*, 1995, **92**, 11949.
4. F. Rappaport and J. Lavergne, *Biochim. Biophys. Acta*, 2001, **1503**, 246.
5. P. R. Rich, B. Meunier, R. Mitchell and A. J. Moody, *Biochim. Biophys. Acta*, 1996, **1275**, 1.
6. M. Wikstrom, *Curr. Opin. Struct. Biol.*, 1998, **8**, 480.
7. R. J. Williams, *Biochim. Biophys. Acta*, 1991, **1058**, 71.
8. B. G. Malmstroem, *Acc. Chem. Res.*, 1993, **26**, 332.
9. J. Stubbe, D. G. Nocera, C. S. Yee and M. C. Chang, *Chem. Rev.*, 2003, **103**, 2167.
10. K. B. Cho, M. A. Carvajal and S. Shaik, *J. Phys. Chem. B*, 2009, **113**, 336.
11. P. Trickey, J. Basran, L. Y. Lian, Z. Chen, J. D. Barton, M. J. Sutcliffe, N. S. Scrutton and F. S. Mathews, *Biochemistry*, 2000, **39**, 7678.
12. R. Camba, Y. S. Jung, L. M. Hunsicker-Wang, B. K. Burgess, C. D. Stout, J. Hirst and F. A. Armstrong, *Biochemistry*, 2003, **42**, 10589.
13. G. A. DiLabio and E. R. Johnson, *J. Am. Chem. Soc.*, 2007, **129**, 6199.

14. L. Noodleman and W. G. Han, *J. Biol. Inorg. Chem.*, 2006, **11**, 674.
15. A. F. Miller, K. Padmakumar, D. L. Sorkin, A. Karapetian and C. K. Vance, *J. Inorg. Biochem.*, 2003, **93**, 71.
16. W. G. Han, T. Lovell and L. Noodleman, *Inorg. Chem.*, 2002, **41**, 205.
17. M. M. Whittaker and J. W. Whittaker, *Biochemistry*, 1997, **36**, 8923.
18. J. M. Mayer, *Annu. Rev. Phys. Chem.*, 2004, **55**, 363.
19. J. M. Mayer and I. J. Rhile, *Biochim. Biophys. Acta*, 2004, **1655**, 51.
20. S. Y. Reece, J. M. Hodgkiss, J. Stubbe and D. G. Nocera, *Philos. Trans. R. Soc., B*, 2006, **361**, 1351.
21. S. Y. Reece and D. G. Nocera, *Annu. Rev. Biochem.*, 2009, **78**, 673.
22. K. Chen, J. Hirst, R. Camba, C. A. Bonagura, C. D. Stout, B. K. Burgess and F. A. Armstrong, *Nature*, 2000, **405**, 814.
23. M. S. Graige, M. L. Paddock, G. Feher and M. Y. Okamura, *Biochemistry*, 1999, **38**, 11465.
24. S. C. Weatherly, I. V. Yang, P. A. Armistead and H. H. Thorp, *J. Phys. Chem. B*, 2003, **107**, 372.
25. M. H. Huynh and T. J. Meyer, *Chem. Rev.*, 2007, **107**, 5004.
26. R. A. Binstead, M. E. McGuire, A. Dovletoglou, W. K. Seok, L. E. Roecker and T. J. Meyer, *J. Am. Chem. Soc.*, 1992, **114**, 173.
27. R. A. Binstead and T. J. Meyer, *J. Am. Chem. Soc.*, 1987, **109**, 3287.
28. R. A. Binstead, B. A. Moyer, G. J. Samuels and T. J. Meyer, *J. Am. Chem. Soc.*, 1981, **103**, 2897.
29. C. Costentin, *Chem. Rev.*, 2008, **108**, 2145.
30. A. Jordan and P. Reichard, *Annu. Rev. Biochem.*, 1998, **67**, 71.
31. W. G. Han, T. Liu, T. Lovell and L. Noodleman, *J. Inorg. Biochem.*, 2006, **100**, 771.
32. B. Katterle, M. Sahlin, P. P. Schmidt, S. Potsch, D. T. Logan, A. Graslund and B. M. Sjoberg, *J. Biol. Chem.*, 1997, **272**, 10414.
33. A. J. Narvaez, N. Voevodskaya, L. Thelander and A. Graslund, *J. Biol. Chem.*, 2006, **281**, 26022.
34. M. R. Seyedsayamdost and J. Stubbe, *J. Am. Chem. Soc.*, 2006, **128**, 2522.
35. J. Stubbe, *Curr. Opin. Chem. Biol.*, 2003, **7**, 183.
36. R. E. Sharp and S. K. Chapman, *Biochim. Biophys. Acta*, 1999, **1432**, 143.
37. P. Brzezinski and R. B. Gennis, *J. Bioenerg. Biomembr.*, 2008, **40**, 521.
38. A. Namslauer, A. S. Pawate, R. B. Gennis and P. Brzezinski, *Proc. Natl. Acad. Sci. U. S. A.*, 2003, **100**, 15543.
39. V. R. Kaila, M. I. Verkhovsky, G. Hummer and M. Wikstrom, *Proc. Natl. Acad. Sci. U. S. A.*, 2008, **105**, 6255.
40. V. R. Kaila, M. Verkhovsky, G. Hummer and M. Wikstrom, *Biochim. Biophys. Acta*, 2008, **1777**, 890.
41. S. A. Siletsky, A. S. Pawate, K. Weiss, R. B. Gennis and A. A. Konstantinov, *J. Biol. Chem.*, 2004, **279**, 52558.
42. A. A. Konstantinov, S. Siletsky, D. Mitchell, A. Kaulen and R. B. Gennis, *Proc. Natl. Acad. Sci. U. S. A.*, 1997, **94**, 9085.
43. M. Oliveberg, P. Brzezinski and B. G. Malmstrom, *Biochim. Biophys. Acta*, 1989, **977**, 322.

44. J. Gebicki, A. Marcinek and J. Zielonka, *Acc. Chem. Res.*, 2004, **37**, 379.
45. J. Yuasa, S. Yamada and S. Fukuzumi, *Angew. Chem., Int. Ed. Engl.*, 2008, **47**, 1068.
46. J. Yuasa, S. Yamada and S. Fukuzumi, *J. Am. Chem. Soc.*, 2008, **130**, 5808.
47. A. Fersht, *Structure and mechanism in protein science: A guide to enzyme catalysis and protein folding*, W.H. Freeman and Company, New York, 1999.
48. C. Bohr, K. Hasselbalch and A. Krogh, *Skand. Arch. Physiol.*, 1904, **16**, 402.
49. F. Q. Schafer and G. R. Buettner, *Free Radic. Biol. Med.*, 2001, **30**, 1191.
50. P. Atkins and J. de Paula, *Atkins' Physical Chemistry*, Oxford University Press, New York, 2002.
51. M. B. Murataliev, R. Feyereisen and F. A. Walker, *Biochim. Biophys. Acta*, 2004, **1698**, 1.
52. S. N. Daff, S. K. Chapman, K. L. Turner, R. A. Holt, S. Govindaraj, T. L. Poulos and A. W. Munro, *Biochemistry*, 1997, **36**, 13816.
53. A. J. Dunford, S. E. Rigby, S. Hay, A. W. Munro and N. S. Scrutton, *Biochemistry*, 2007, **46**, 5018.
54. A. W. Munro, M. A. Noble, L. Robledo, S. N. Daff and S. K. Chapman, *Biochemistry*, 2001, **40**, 1956.
55. K. R. Wolthers, J. Basran, A. W. Munro and N. S. Scrutton, *Biochemistry*, 2003, **42**, 3911.
56. T. Iyanagi, N. Makino and H. S. Mason, *Biochemistry*, 1974, **13**, 1701.
57. S. Ghisla, V. Massey, J. M. Lhoste and S. G. Mayhew, *Biochemistry*, 1974, **13**, 589.
58. Y. T. Kao, C. Saxena, T. F. He, L. Guo, L. Wang, A. Sancar and D. Zhong, *J. Am. Chem. Soc.*, 2008, **130**, 13132.
59. G. N. Yalloway, S. G. Mayhew, J. P. Malthouse, M. E. Gallagher and G. P. Curley, *Biochemistry*, 1999, **38**, 3753.
60. S. G. Mayhew, *Eur. J. Biochem.*, 1999, **265**, 698.
61. M. L. Ludwig, L. M. Schopfer, A. L. Metzger, K. A. Pattridge and V. Massey, *Biochemistry*, 1990, **29**, 10364.
62. F. A. Armstrong, *Philos. Trans. R. Soc., B*, 2008, **363**, 1263.
63. J. Barber, *Biochim. Biophys. Acta*, 2004, **1655**, 123.
64. H. Dau, L. Iuzzolino and J. Dittmer, *Biochim. Biophys. Acta*, 2001, **1503**, 24.
65. H. Dau and M. Haumann, *Biochim. Biophys. Acta*, 2007, **1767**, 472.
66. H. Dau and M. Haumann, *Photosynth. Res.*, 2005, **84**, 325.
67. H. Dau, P. Liebisch and M. Haumann, *Anal. Bioanal. Chem.*, 2003, **376**, 562.
68. S. Hotchandani, U. Ozdemir, C. Nasr, S. I. Allakhverdiev, N. Karacan, V. V. Klimov, P. V. Kamat and R. Carpentier, *Bioelectrochem. Bioenerg.*, 1999, **48**, 53.
69. W. Hillier and T. Wydrzynski, *Biochim. Biophys. Acta*, 2001, **1503**, 197.
70. C. I. Lee, K. V. Lakshmi and G. W. Brudvig, *Biochemistry*, 2007, **46**, 3211.

71. L. P. Vincent, M. J. Baldwin, M. T. Caudle, W. Hsieh and N. A. Law, *Pure Appl. Chem.*, 1998, **70**, 925.
72. S. Brenner, S. Hay, A. W. Munro and N. S. Scrutton, *FEBS J.*, 2008, **275**, 4540.
73. M. Pourbaix, in *Atlas of Electrochemical Equilibria in Aqueous Solution*, ed. J. A. Franklin, Pergamon Press, New York, 1974.
74. P. L. Dutton, *Methods Enzymol.*, 1978, **54**, 411.
75. K. J. McLean, N. S. Scrutton and A. W. Munro, *Biochem. J.*, 2003, **372**, 317.
76. A. W. Munro, S. N. Daff, K. L. Turner and S. K. Chapman, *Biochem. Soc. Trans.*, 1997, **25**, S628.
77. E. J. Dridge, C. A. Watts, B. J. Jepson, K. Line, J. M. Santini, D. J. Richardson and C. S. Butler, *Biochem. J.*, 2007, **408**, 19.
78. L. Euro, D. A. Bloch, M. Wikstrom, M. I. Verkhovsky and M. Verkhovskaya, *Biochemistry*, 2008, **47**, 3185.
79. C. R. Staples, I. K. Dhawan, M. G. Finnegan, D. A. Dwinell, Z. H. Zhou, H. Huang, M. F. Verhagen, M. W. Adams and M. K. Johnson, *Inorg. Chem.*, 1997, **36**, 5740.
80. F. A. Armstrong, J. N. Butt and A. Sucheta, *Methods Enzymol.*, 1993, **227**, 479.
81. C. Leger, S. J. Elliott, K. R. Hoke, L. J. Jeuken, A. K. Jones and F. A. Armstrong, *Biochemistry*, 2003, **42**, 8653.
82. F. A. Armstrong, *Curr. Opin. Chem. Biol.*, 2005, **9**, 110.
83. H. B. Gray and B. G. Malmstrom, *Biochemistry*, 1989, **28**, 7499.
84. H. B. Gray and J. R. Winkler, *Annu. Rev. Biochem.*, 1996, **65**, 537.
85. H. B. Gray and J. R. Winkler, *Q. Rev. Biophys.*, 2003, **36**, 341.
86. R. A. Marcus, *J. Chem. Phys.*, 2006, **125**, 194504.
87. R. A. Marcus, *Philos. Trans. R. Soc., B*, 2006, **361**, 1445.
88. R. A. Marcus and N. Sutin, *Biochim. Biophys. Acta*, 1985, **811**, 265.
89. C. C. Moser, J. M. Keske, K. Warncke, R. S. Farid and P. L. Dutton, *Nature*, 1992, **355**, 796.
90. R. I. Cukier, *J. Phys. Chem.*, 1995, **99**, 16101.
91. R. I. Cukier, *J. Phys. Chem.*, 1994, **98**, 2377.
92. R. I. Cukier, *J. Phys. Chem.*, 1996, **100**, 15428.
93. S. Hammes-Schiffer, *Acc. Chem. Res.*, 2001, **34**, 273.
94. S. Hammes-Schiffer, *ChemPhysChem*, 2002, **3**, 33.
95. S. Hammes-Schiffer, *Acc. Chem. Res.*, 2006, **39**, 93.
96. S. Hammes-Schiffer and N. Iordanova, *Biochim. Biophys. Acta*, 2004, **1655**, 29.
97. S. Hammes-Schiffer and A. V. Soudackov, *J. Phys. Chem. B*, 2008, **112**, 14108.
98. X. G. Zhao and R. I. Cukier, *J. Phys. Chem.*, 1995, **99**, 945.
99. P. L. Dutton and C. C. Mosser, *Proc. Natl. Acad. Sci. U. S. A.*, 1994, **91**, 10247.
100. V. L. Davidson, *Acc. Chem. Res.*, 2000, **33**, 87.
101. J. J. Hopfield, *Proc. Natl. Acad. Sci. U. S. A.*, 1974, **71**, 3640.

102. J. N. Onuchic, D. N. Beratan, J. R. Winkler and H. B. Gray, *Annu. Rev. Biophys. Biomol. Struct.*, 1992, **21**, 349.
103. G. W. Canters and C. Dennison, *Biochimie*, 1995, **77**, 506.
104. A. M. Ponce and L. E. Overman, *J. Am. Chem. Soc.*, 2000, **122**, 8672.
105. D. N. Beratan, J. N. Betts and J. N. Onuchic, *Science*, 1991, **252**, 1285.
106. V. L. Davidson, *Biochemistry*, 1996, **35**, 14035.
107. V. G. Levich, *Adv. Electrochem. Electrochem. Eng.*, 1966, **4**, 249.
108. V. G. Levich and R. R. Dogonadze, *Dokl. Acad. Nauk. USSR*, 1959, **124**, 123.
109. V. L. Davidson, *Biochemistry*, 2002, **41**, 14633.
110. V. L. Davidson, *Biochemistry*, 2000, **39**, 4924.
111. T. K. Harris, V. L. Davidson, L. Chen, F. S. Mathews and Z. X. Xia, *Biochemistry*, 1994, **33**, 12600.
112. J. R. Winkler and H. B. Gray, *Chem. Rev.*, 1992, **92**, 369.
113. B. M. Hoffman and M. A. Ratner, *J. Am. Chem. Soc.*, 1987, **109**, 6237.
114. H. J. Lee, J. Basran and N. S. Scrutton, *Biochemistry*, 1998, **37**, 15513.
115. V. L. Davidson, *Arch. Biochem. Biophys.*, 2004, **428**, 32.
116. M. J. Knapp and J. P. Klinman, *Eur. J. Biochem*, 2002, **269**, 3113.
117. J. P. Klinman, *Biochim. Biophys. Acta*, 2006, **1757**, 981.
118. S. Hay and N. S. Scrutton, *Photosynth. Res.*, 2008, **98**, 169.
119. J. H. Skone, A. V. Soudackov and S. Hammes-Schiffer, *J. Am. Chem. Soc.*, 2006, **128**, 16655.
120. A. Soudackov and S. Hammes-Schiffer, *J. Chem. Phys.*, 2000, **113**, 2385.
121. S. Hammes-Schiffer and S. J. Benkovic, *Annu. Rev. Biochem.*, 2006, **75**, 519.
122. S. J. Benkovic and S. Hammes-Schiffer, *Science*, 2006, **312**, 208.
123. S. J. Benkovic and S. Hammes-Schiffer, *Science*, 2003, **301**, 1196.
124. A. M. Kuznetsov and J. Ulstrup, *Can. J. Chem.*, 1999, **77**, 1085.
125. E. P. Friis, J. E. Andersen, Y. I. Kharkats, A. M. Kuznetsov, R. J. Nichols, J. D. Zhang and J. Ulstrup, *Proc. Natl. Acad. Sci. U. S. A.*, 1999, **96**, 1379.
126. J. P. Klinman, *Philos. Trans. R. Soc., B*, 2006, **361**, 1323.
127. J. Basran, M. J. Sutcliffe and N. S. Scrutton, *Biochemistry*, 1999, **38**, 3218.
128. N. S. Scrutton, J. Basran and M. J. Sutcliffe, *Eur. J. Biochem.*, 1999, **264**, 666.
129. A. Kohen and J. P. Klinman, *Chem. Biol.*, 1999, **6**, R191.
130. A. Kohen, R. Cannio, S. Bartolucci and J. P. Klinman, *Nature*, 1999, **399**, 496.
131. M. J. Sutcliffe, L. Masgrau, A. Roujeinikova, L. O. Johannissen, P. Hothi, J. Basran, K. E. Ranaghan, A. J. Mulholland, D. Leys and N. S. Scrutton, *Philos. Trans. R. Soc., B*, 2006, **361**, 1375.
132. S. Hay, C. Pudney, P. Hothi, L. O. Johannissen, L. Masgrau, J. Pang, D. Leys, M. J. Sutcliffe and N. S. Scrutton, *Biochem. Soc. Trans.*, 2008, **36**, 16.
133. R. K. Allemann, R. M. Evans, L. H. Tey, G. Maglia, J. Y. Pang and R. Rodriguez, *Philos. Trans. R. Soc. B*, 2006, **361**, 1317.

134. L. Wang, N. M. Goodey, S. J. Benkovic and A. Kohen, *Philos. Trans. R. Soc. B*, 2006, **361**, 1307.
135. D. B. Northrop, ed. W. W. Cleland, M. H. O'Leary and D. B. Northrop, *Isotope Effects on Enzyme-Catalyzed Reactions*, University Park Press, Baltimore, 1977, pp. 122.
136. D. B. Northrop, *Annu. Rev. Biochem.*, 1981, **50**, 103.
137. K. B. Schowen and R. L. Schowen, *Methods Enzymol.*, 1982, **87**, 551.
138. B. Hong, F. Maley and A. Kohen, *Biochemistry*, 2007, **46**, 14188.
139. B. Hong, M. Haddad, F. Maley, J. H. Jensen and A. Kohen, *J. Am. Chem. Soc.*, 2006, **128**, 5636.
140. J. Basran, R. J. Harris, M. J. Sutcliffe and N. S. Scrutton, *J. Biol. Chem.*, 2003, **278**, 43973.
141. C. R. Pudney, S. Hay, J. Pang, C. Costello, D. Leys, M. J. Sutcliffe and N. S. Scrutton, *J. Am. Chem. Soc.*, 2007, **129**, 13949.
142. C. R. Pudney, S. Hay, M. J. Sutcliffe and N. S. Scrutton, *J. Am. Chem. Soc.*, 2006, **128**, 14053.
143. M. M. Purdy, L. S. Koo, P. R. de Montellano and J. P. Klinman, *Biochemistry*, 2006, **45**, 15793.
144. S. C. Sharma and J. P. Klinman, *J. Am. Chem. Soc.*, 2008, **130**, 17632.
145. J. P. Evans, N. J. Blackburn and J. P. Klinman, *Biochemistry*, 2006, **45**, 15419.
146. K. Takahashi and J. P. Klinman, *Biochemistry*, 2006, **45**, 4683.
147. H. S. Kim, S. M. Damo, S. Y. Lee, D. Wemmer and J. P. Klinman, *Biochemistry*, 2005, **44**, 11428.
148. K. M. Doll and R. G. Finke, *Inorg. Chem.*, 2003, **42**, 4849.
149. E. N. Segraves and T. R. Holman, *Biochemistry*, 2003, **42**, 5236.
150. W. A. Francisco, N. J. Blackburn and J. P. Klinman, *Biochemistry*, 2003, **42**, 1813.
151. S. L. Seymour and J. P. Klinman, *Biochemistry*, 2002, **41**, 8747.
152. S. Tsai and J. P. Klinman, *Biochemistry*, 2001, **40**, 2303.
153. K. W. Rickert and J. P. Klinman, *Biochemistry*, 1999, **38**, 12218.
154. K. Takahashi, T. Onami and M. Noguchi, *Biochem. J.*, 1998, **336**, 131.
155. W. A. Francisco, D. J. Merkler, N. J. Blackburn and J. P. Klinman, *Biochemistry*, 1998, **37**, 8244.
156. D. Antoniou and S. D. Schwartz, *Proc. Natl. Acad. Sci. U. S. A.*, 1997, **94**, 12360.
157. A. Kohen, T. Jonsson and J. P. Klinman, *Biochemistry*, 1997, **36**, 2603.
158. M. H. Glickman and J. P. Klinman, *Biochemistry*, 1995, **34**, 14077.
159. T. Jonsson, D. E. Edmondson and J. P. Klinman, *Biochemistry*, 1994, **33**, 14871.
160. M. M. Palcic and J. P. Klinman, *Biochemistry*, 1983, **22**, 5957.
161. K. M. Welsh, D. J. Creighton and J. P. Klinman, *Biochemistry*, 1980, **19**, 2005.
162. J. Pu, S. Ma, M. Garcia-Viloca, J. Gao, D. G. Truhlar and A. Kohen, *J. Am. Chem. Soc.*, 2005, **127**, 14879.

163. N. Agrawal, S. A. Lesley, P. Kuhn and A. Kohen, *Biochemistry*, 2004, **43**, 10295.
164. Q. Su and J. P. Klinman, *Biochemistry*, 1998, **37**, 12513.
165. M. M. Whittaker, D. P. Ballou and J. W. Whittaker, *Biochemistry*, 1998, **37**, 8426.
166. R. V. Dunn, K. R. Marshall, A. W. Munro and N. S. Scrutton, *FEBS J.*, 2008, **275**, 3850.
167. D. J. Heyes, M. Sakuma and N. S. Scrutton, *Angew. Chem., Int. Ed. Engl.*, 2009, **48**, 3850.
168. J. Basran, N. Bhanji, A. Basran, D. Nietlispach, S. Mistry, R. Meskys and N. S. Scrutton, *Biochemistry*, 2002, **41**, 4733.
169. J. Basran, M. H. Jang, M. J. Sutcliffe, R. Hille and N. S. Scrutton, *J. Biol. Chem.*, 1999, **274**, 13155.
170. J. Basran, S. Patel, M. J. Sutcliffe and N. S. Scrutton, *J. Biol. Chem.*, 2001, **276**, 6234.
171. J. Basran, M. J. Sutcliffe, R. Hille and N. S. Scrutton, *Biochem. J.*, 1999, **341**, 307.
172. J. Basran, M. J. Sutcliffe and N. S. Scrutton, *J. Biol. Chem.*, 2001, **276**, 24581.
173. R. J. Harris, R. Meskys, M. J. Sutcliffe and N. S. Scrutton, *Biochemistry*, 2000, **39**, 1189.
174. P. Hothi, J. Basran, M. J. Sutcliffe and N. S. Scrutton, *Biochemistry*, 2003, **42**, 3966.
175. P. Hothi, K. A. Khadra, J. P. Combe, D. Leys and N. S. Scrutton, *FEBS J.*, 2005, **272**, 5894.
176. P. Hothi, M. J. Sutcliffe and N. S. Scrutton, *Biochem. J.*, 2005, **388**, 123.
177. F. Polticelli, J. Basran, C. Faso, A. Cona, G. Minervini, R. Angelini, R. Federico, N. S. Scrutton and P. Tavladoraki, *Biochemistry*, 2005, **44**, 16108.
178. K. R. Wolthers and N. S. Scrutton, *Biochemistry*, 2004, **43**, 490.
179. S. Hay, J. Pang, P. J. Monaghan, X. Wang, R. M. Evans, M. J. Sutcliffe, R. K. Allemann and N. S. Scrutton, *ChemPhysChem*, 2008, **9**, 1536.
180. S. Hay, C. R. Pudney, M. J. Sutcliffe and N. S. Scrutton, *ChemPhysChem*, 2008, **9**, 1875.
181. S. Hay, M. J. Sutcliffe and N. S. Scrutton, *Proc. Natl. Acad. Sci. U. S. A.*, 2007, **104**, 507.
182. P. Hothi, S. Hay, A. Roujeinikova, M. J. Sutcliffe, M. Lee, D. Leys, P. M. Cullis and N. S. Scrutton, *ChemBioChem*, 2008, **9**, 2839.
183. D. L. Jenson, A. Evans and B. A. Barry, *J. Phys. Chem. B*, 2007, **111**, 12599.
184. D. B. Northrop, *Biochemistry*, 1975, **14**, 2644.
185. Y. Cha, C. J. Murray and J. P. Klinman, *Science*, 1989, **243**, 1325.
186. C. G. Swain, E. C. Stivers, J. F. Reuwer and L. J. Schaad, *J. Am. Chem. Soc.*, 1958, **80**, 5885.

187. C. Costentin, M. Robert and J. M. Saveant, *J. Am. Chem. Soc.*, 2007, **129**, 5870.
188. P. Adelroth, M. S. Ek, D. M. Mitchell, R. B. Gennis and P. Brzezinski, *Biochemistry*, 1997, **36**, 13824.
189. M. Svensson-Ek, J. W. Thomas, R. B. Gennis, T. Nilsson and P. Brzezinski, *Biochemistry*, 1996, **35**, 13673.
190. K. Faxen, G. Gilderson, P. Adelroth and P. Brzezinski, *Nature*, 2005, **437**, 286.
191. S. Paula, A. Sucheta, I. Szundi and O. Einarsdottir, *Biochemistry*, 1999, **38**, 3025.
192. M. Svensson, S. Hallen, J. W. Thomas, L. J. Lemieux, R. B. Gennis and T. Nilsson, *Biochemistry*, 1995, **34**, 5252.
193. J. Sasaki and J. L. Spudich, *Biophys. J.*, 1999, **77**, 2145.
194. A. R. Jones, S. Hay, J. R. Woodward and N. S. Scrutton, *J. Am. Chem. Soc.*, 2007, **129**, 15718.
195. A. R. Jones, N. S. Scrutton and J. R. Woodward, *J. Am. Chem. Soc.*, 2006, **128**, 8408.
196. J. R. Woodward, T. J. Foster, A. R. Jones, A. T. Salaoru and N. S. Scrutton, *Biochem. Soc. Trans.*, 2009, **37**, 358.
197. M. L. Verkhovskaya, N. Belevich, L. Euro, M. Wikstrom and M. I. Verkhovsky, *Proc. Natl. Acad. Sci. U. S. A.*, 2008, **105**, 3763.
198. A. Y. Semenov, I. R. Vassiliev, A. van Der Est, M. D. Mamedov, B. Zybailov, G. Shen, D. Stehlik, B. A. Diner, P. R. Chitnis and J. H. Golbeck, *J. Biol. Chem.*, 2000, **275**, 23429.
199. M. Di Valentin, A. Bisol, G. Agostini and D. Carbonera, *J. Chem. Inf. Model*, 2005, **45**, 1580.
200. A. Van Der Est, A. I. Valieva, Y. E. Kandrashkin, G. Shen, D. A. Bryant and J. H. Golbeck, *Biochemistry*, 2004, **43**, 1264.
201. W. Xu, P. R. Chitnis, A. Valieva, A. van der Est, K. Brettel, M. Guergova-Kuras, Y. N. Pushkar, S. G. Zech, D. Stehlik, G. Shen, B. Zybailov and J. H. Golbeck, *J. Biol. Chem.*, 2003, **278**, 27876.
202. F. A. Armstrong, *J. Chem. Soc., Dalton Trans.*, 2002, 661 .
203. M. G. Almeida, C. M. Silveira, B. Guigliarelli, P. Bertrand, J. J. Moura, I. Moura and C. Leger, *FEBS Lett.*, 2007, **581**, 284.
204. B. A. Averill, *Chem. Rev.*, 1996, **96**, 2951.
205. H. Ichiki, Y. Tanaka, K. Mochizuki, K. Yoshimatsu, T. Sakurai and T. Fujiwara, *J. Bacteriol.*, 2001, **183**, 4149.
206. W. G. Zumft, *Microbiol. Mol. Biol. Rev.*, 1997, **61**, 533.
207. S. Suzuki, K. Kataoka, K. Yamaguchi, T. Inoue and Y. Kai, *Coord. Chem. Rev.*, 1999, **190–192**, 245.
208. S. Suzuki, K. Kataoka and K. Yamaguchi, *Acc. Chem. Res.*, 2000, **33**, 728.
209. B. D. Howes, Z. H. Abraham, D. J. Lowe, T. Bruser, R. R. Eady and B. E. Smith, *Biochemistry*, 1994, **33**, 3171.
210. F. E. Dodd, S. S. Hasnain, Z. H. Abraham, R. R. Eady and B. E. Smith, *Acta Crystallogr., D Biol. Crystallogr.*, 1997, **53**, 406.

211. M. J. Ellis, M. Prudencio, F. E. Dodd, R. W. Strange, G. Sawers, R. R. Eady and S. S. Hasnain, *J. Mol. Biol.*, 2002, **316**, 51.
212. S. A. De Marothy, M. R. Blomberg and P. E. Siegbahn, *J. Comput. Chem.*, 2007, **28**, 528.
213. M. Sundararajan, I. H. Hillier and N. A. Burton, *J. Phys. Chem. B*, 2007, **111**, 5511.
214. S. V. Antonyuk, R. W. Strange, G. Sawers, R. R. Eady and S. S. Hasnain, *Proc. Natl. Acad. Sci. U. S. A.*, 2005, **102**, 12041.
215. K. Kataoka, H. Furusawa, K. Takagi, K. Yamaguchi and S. Suzuki, *J. Biochem.*, 2000, **127**, 345.
216. K. Kobayashi, S. Tagawa, Deligeer and S. Suzuki, *J. Biochem.*, 1999, **126**, 408.
217. I. S. MacPherson and M. E. Murphy, *Cell. Mol. Life Sci.*, 2007, **64**, 2887.
218. Z. H. Abraham, B. E. Smith, B. D. Howes, D. J. Lowe and R. R. Eady, *Biochem. J.*, 1997, **324**, 511.
219. O. Farver, R. R. Eady, Z. H. Abraham and I. Pecht, *FEBS Lett.*, 1998, **436**, 239.
220. O. Farver, R. R. Eady, G. Sawers, M. Prudencio and I. Pecht, *FEBS Lett.*, 2004, **561**, 173.
221. S. Suzuki, Deligeer, K. Yamaguchi, K. Kataoka, K. Kobayashi, S. Tagawa, T. Kohzuma, S. Shidara and H. Iwasaki, *J. Biol. Inorg. Chem.*, 1997, **2**, 265.
222. S. Brenner, D. J. Heyes, S. Hay, M. A. Hough, R. R. Eady, S. S. Hasnain and N. S. Scrutton, *J. Biol. Chem.*, 2009, **284**, 25973.
223. R. W. Strange, L. M. Murphy, F. E. Dodd, Z. H. Abraham, R. R. Eady, B. E. Smith and S. S. Hasnain, *J. Mol. Biol.*, 1999, **287**, 1001.
224. D. H. Craig, P. C. Moody, N. C. Bruce and N. S. Scrutton, *Biochemistry*, 1998, **37**, 7598.
225. S. Hay, C. R. Pudney, P. Hothi and N. S. Scrutton, *J. Phys. Chem. A*, 2008.
226. H. Khan, R. J. Harris, T. Barna, D. H. Craig, N. C. Bruce, A. W. Munro, P. C. Moody and N. S. Scrutton, *J. Biol. Chem.*, 2002, **277**, 21906.
227. J. Pang, S. Hay, N. S. Scrutton and M. J. Sutcliffe, *J. Am. Chem. Soc.*, 2008, **130**, 7092.
228. J. Basran, R. J. Harris, M. J. Sutcliffe and N. S. Scrutton, *J. Biol. Chem.*, 2003, **278**, 43973.
229. S. Hay, C. R. Pudney, P. Hothi and N. S. Scrutton, *J. Phys. Chem. A*, 2008, **112**, 13109.
230. W. J. Bruno and W. Bialek, *Biophys. J.*, 1992, **63**, 689.
231. M. J. Knapp, K. Rickert and J. P. Klinman, *J. Am. Chem. Soc.*, 2002, **124**, 3865.
232. S. Hay, M. J. Sutcliffe and N. S. Scrutton, in *Quantum Tunnelling in Enzyme Catalyzed Reactions*, ed. R. K. Allemann and N. S. Scrutton, Royal Society of Chemistry, Cambridge, 2009, pp. 199.
233. S. Hay, C. R. Pudney, T. A. McGrory, J. Pang, M. J. Sutcliffe and N. S. Scrutton, *Angew. Chem., Int. Ed. Engl.*, 2009, **48**, 1452.

234. S. J. Benkovic, G. G. Hammes and S. Hammes-Schiffer, *Biochemistry*, 2008, **47**, 3317.
235. C. R. Pudney, T. McGrory, P. Lafite, J. Pang, S. Hay, D. Leys, M. J. Sutcliffe and N. S. Scrutton, *ChemBioChem*, 2009, **10**, 1379.
236. M. Sjodin, T. Irebo, J. E. Utas, J. Lind, G. Merenyi, B. Akermark and L. Hammarstrom, *J. Am. Chem. Soc.*, 2006, **128**, 13076.
237. T. J. Meyer, M. H. Huynh and H. H. Thorp, *Angew. Chem., Int. Ed. Engl.*, 2007, **46**, 5284.
238. C. Costentin, D. H. Evans, M. Robert, J. M. Saveant and P. S. Singh, *J. Am. Chem. Soc.*, 2005, **127**, 12490.
239. I. J. Rhile, T. F. Markle, H. Nagao, A. G. DiPasquale, O. P. Lam, M. A. Lockwood, K. Rotter and J. M. Mayer, *J. Am. Chem. Soc.*, 2006, **128**, 6075.
240. R. I. Cukier, *J. Phys. Chem. B*, 2002, **106**, 1746.
241. R. I. Cukier and D. G. Nocera, *Annu. Rev. Phys. Chem.*, 1998, **49**, 337.
242. W. W. Cleland, *CRC Crit. Rev. Biochem.*, 1982, **13**, 385.
243. R. Bell, *The Tunnel Effect in Chemistry*, Chapman and Hall, London, 1980.
244. D. Borgis and J. T. Hynes, *J. Phys. Chem.*, 1996, **100**, 1118.
245. L. Masgrau, J. Basran, P. Hothi, M. J. Sutcliffe and N. S. Scrutton, *Arch. Biochem. Biophys.*, 2004, **428**, 41.

Metal Ion-Coupled and Proton-Coupled Electron Transfer in Catalytic Reduction of Dioxygen

SHUNICHI FUKUZUMI*[a,b] AND HIROAKI KOTANI[a]

[a] Department of Material and Life Science, Graduate School of Engineering, Osaka University, ALCA, Japan Science and Technology Agency (JST), Suita, Osaka 565-0871, Japan; [b] Department of Bioinspired Science, Ewha Womans University, Seoul 120-750, Korea

4.1 Introduction

Dioxygen (O_2) is the most important oxidant in the respiration of mammals and in oxidative metabolic processes.[1,2] O_2 is triplet in the ground state and thereby any reaction with O_2 to afford singlet oxygenated products is spin-forbidden.[3,4] Such limited reactivity of O_2 due to the spin-forbidden reaction is essential for maintaining life, because otherwise O_2 would burn the substrates to CO_2 rather than supply energy in a controlled way.[4] The only spin-allowed reaction of the ground state O_2 is electron transfer from electron donors to O_2.[5-7] However, the electron transfer to O_2 requires strong electron donors because the one-electron reduction potential of O_2 is quite negative.[8] The unfavorable energetics of the electron-transfer reduction can be altered to be favorable by the binding of a proton or metal ions to superoxide anion ($O_2^{\cdot-}$), because the binding of protons and metal ions to radical anions of electron acceptors results in significant positive shifts of the one-electron reduction

RSC Catalysis Series No. 8
Proton-Coupled Electron Transfer: A Carrefour of Chemical Reactivity Traditions
Edited by Sebastião Formosinho and Mónica Barroso
© Royal Society of Chemistry 2012
Published by the Royal Society of Chemistry, www.rsc.org

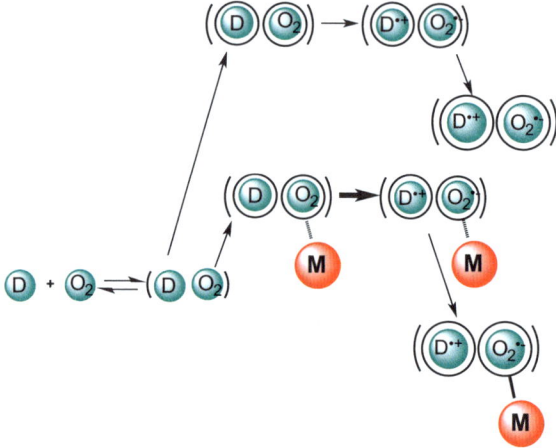

Scheme 4.1   Schematic energy diagram for PCET and MCET from an electron donor
(D) to O_2 with and without M (M = H^+ and metal ions for PCET and
MCET, respectively).

potentials of electron acceptors.[9,10] Thus, a variety of uphill electron-transfer
reactions, which are thermodynamically infeasible, are made possible by the
presence of an acid or metal ions, provided that the strong binding of an acid or
metal ions to radical anions of electron acceptors change the energetics of
electron transfer from uphill to downhill, as shown in Scheme 4.1 for the case of
O_2.[5,9–18] In this case, protons and metal ions can promote electron-transfer
reactions that would otherwise be impossible, which are defined as proton-
coupled electron transfer (PCET) and metal ion-coupled electron transfer
(MCET).[19–25]

   PCET and MCET play pivotal roles in biological electron-transfer systems,
such as photosynthesis and respiration.[25–30] The highly exergonic four-electron
reduction of O_2 to water, which is the reverse process of photosynthesis,
maintains the life of aerobic organisms by respiration.[29,30] Cytochrome c oxidases (CcOs) are responsible for catalyzing the reduction of O_2 to water by the
soluble electron carrier, cytochrome c.[29,30] The X-ray structures of CcOs have
revealed that the catalytic site of CcOs consists of the bimetallic complex of
heme and Cu, where the distance between Fe and Cu has been reported as 4.5 Å
in the absence of O_2.[31] A number of synthetic analogs of the CcO active site
have been synthesized to mimic the coordination environment of the Fe/Cu
core as well as the catalytic function.[32–34] The four-electron reduction of O_2 is
not only of great biological interest, but also of technological significance for
applications such as fuel cells.[33–37]

   The most important question is how the CcO enzyme catalyzes the four-
electron reduction of O_2 to water without releasing the two-electron reduced
species (H_2O_2). This chapter focuses on the MCET and PCET mechanisms of
the one-electron, two-electron, and four-electron reduction of O_2 in homogeneous solutions.

4.2 PCET from Electron Donors to O_2

The one-electron reduction potential (E_{red}) of O_2 is shifted in the positive direction by the protonation of the one-electron reduced product, *i.e.*, superoxide anion $O_2^{\bullet-}$ (eqn (1)), according to the Nernst equation (eqn (2)),

$$HO_2^{\bullet} \xrightarrow{k_a} O_2^{\bullet-} + H^+ \tag{1}$$

$$E_{red} = E^0(O_2/O_2^{\bullet-}) + (2.3RT/F)(\log[H^+] - \log K_a) \tag{2}$$

where $E^0(O_2/O_2^{\bullet-})$ is the one-electron reduction potential of O_2 without an acid, F is the Faraday constant and $2.3RT/F$ corresponds to 59 mV at 298 K, and K_a is the acid dissociation constant of $HO_2^{\bullet}$. From the $E^0(O_2/O_2^{\bullet-})$ value[38] and pK_a ($=12$) in an aprotic polar solvent,[8] the one-electron reduction potential of O_2 in the presence of $HClO_4$ (0.10 M) in acetonitrile (MeCN) is determined from eqn (2) to be –0.21 V.[39]

The rate constant (k_{et}) of outer-sphere electron transfer from an electron donor (D) to an electron acceptor (A) can be estimated by the Marcus equation (eqn (3)),

$$k_{et} = (k_{exD}k_{exA}K_{et}f)^{1/2} \tag{3}$$

where k_{exD} and k_{exA} are the rate constants of the electron self-exchange between D and $D^{\bullet+}$ and that between $A^{\bullet-}$ and A, respectively and K_{et} is the electron-transfer equilibrium.[40] The K_{et} value is obtained from the one-electron oxidation potential of D (E_{ox}) and the one-electron reduction potential of A by eqn (4). The parameter f in eqn (3) is given by eqn (5), where Z is the frequency factor, which is normally taken as 1×10^{11} M^{-1} s^{-1}. In the case of slow electron-transfer reactions, the f value can be approximated to be unity.

$$\log K_{et} = (-2.3RT/F)^{-1}(E_{ox} - E_{red}) \tag{4}$$

$$\log f = (\log K_{et})^2 / [4\log(k_{etD}k_{etA}/Z^2)] \tag{5}$$

The k_{exD} value of 1,1'-dimethylferrocene (Me_2Fc) has been reported to be 8.3×10^6 M^{-1} s^{-1}.[41] Although the k_{exA} value of O_2 is known to vary depending on the type of electron-transfer reactions, in particular because of contribution of an inner-sphere pathway,[42] the most appropriate k_{exA} value was evaluated from outer-sphere electron transfer from $O_2^{\bullet-}$ to ferrocenium cation to be 6.5×10^{-5} M^{-1} s^{-1}.[42] Based on these k_{exD} and k_{exA} values and the K_{et} value, the rate constant of electron transfer from Me_2Fc to O_2 in the presence of $HClO_4$ (0.10 M) in MeCN at 298 K can be estimated to be 1.0×10^{-3} M^{-1} s^{-1}.[39]

No electron transfer occurred from Me_2Fc to O_2 in MeCN because the electron transfer is highly endergonic judging from the E_{ox} value of Me_2Fc (0.28 V *versus* SCE)[43] and the E_{red} value of O_2 (–0.86 V *versus* SCE).[38] In the presence of $HClO_4$ (0.10 M), however, electron transfer from Me_2Fc to O_2

occurs to yield Me_2Fc^+ and $HO_2{}^\bullet$ (eqn (6)).[39] The $HO_2{}^\bullet$ is rapidly reduced by Me_2Fc to produce H_2O_2 and Me_2Fc^+ (eqn (7)).[39]

$$Me_2Fc + O_2 + H^+ \xrightarrow{k_{et}} Me_2Fc^+ + HO_2{}^\bullet \qquad (6)$$

$$Me_2Fc + HO_2{}^\bullet + H^+ \xrightarrow{fast} Me_2Fc^+ + H_2O_2 \qquad (7)$$

The rate-determining step is the initial PCET from Me_2Fc to O_2 and the k_{et} value of PCET was determined to be 1.2×10^{-3} M^{-1} s^{-1}, which agrees with the predicted value (1.0×10^{-3} M^{-1} s^{-1}) predicted from the Marcus equation (eqn (4)).[39] The k_{et} value increased linearly with increasing concentration of $HClO_4$ (Figure 4.1).[39]

Me_2Fc, which is a one-electron reductant, can be replaced by 10-methyl-9,10-dihydroacridine ($AcrH_2$), which is a two-electron reductant in the PCET to O_2.[39] No reduction of O_2 occurs with $AcrH_2$ in MeCN at 298 K. However, the addition of $HClO_4$ to an oxygen-saturated MeCN solution of $AcrH_2$ resulted in the two-electron reduction of O_2 by $AcrH_2$ to yield 10-methylacridinium ion ($AcrH^+$) and H_2O_2 (eqn (8)).[39]

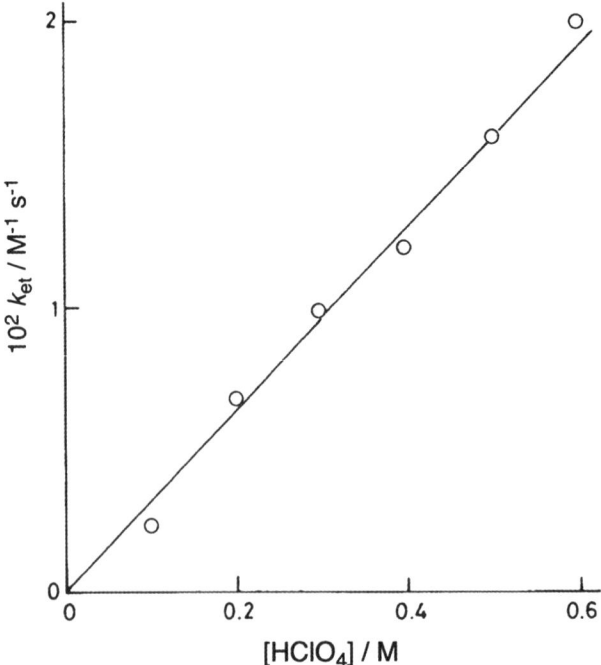

Figure 4.1   Dependence of k_{et} of PCET from Me_2Fc to O_2 in MeCN at 298 K on concentration of $HClO_4$.

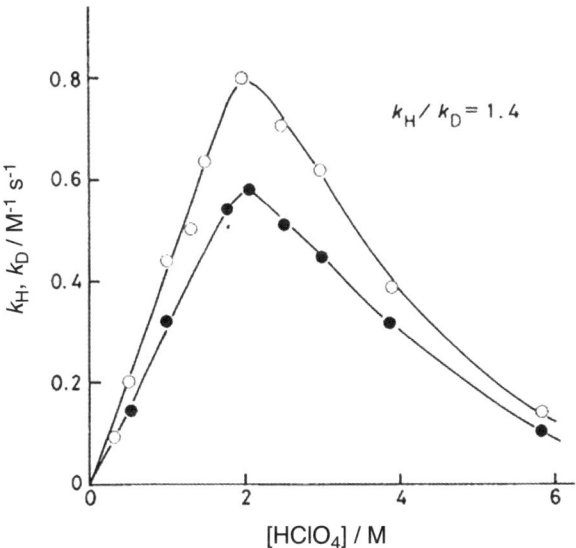

$$\text{AcrH}_2 + O_2 + H^+ \longrightarrow \text{AcrH}^+ + H_2O_2 \tag{8}$$

When a 1:1 (v/v) mixture of MeCN and H_2O was used as a solvent, the two-electron reduction of O_2 by AcrH$_2$ in the presence of HClO$_4$ occurred more efficiently.[39] The rate of formation of AcrH$^+$ obeyed pseudo-first-order kinetics in the presence of excess O_2 and HClO$_4$, and the pseudo-first-order rate constant was proportional to the concentration of O_2 (eqn (9)).[39]

$$d[\text{AcrH}^+]/dt = k_H[\text{AcrH}_2][O_2] \tag{9}$$

The rate of formation of AcrH$^+$ obeyed pseudo-first-order kinetics in the presence of excess O_2 and HClO$_4$.[39] The dependence of the second-order rate constant (k_H) on the HClO$_4$ concentration in an aqueous solution is shown in Figure 4.2.[39] When AcrH$_2$ is replaced by the 9,9'-dideuterated analog (AcrD$_2$), a small primary kinetic isotope effect ($k_H/k_D = 1.4 \pm 0.2$) was observed and the k_H/k_D value remains constant with the variation in the HClO$_4$ concentration (Figure 4.2).[39] The k_H and k_D values increase with an increase in the HClO$_4$ concentration but decrease at high HClO$_4$ concentrations, with a maximum at

Figure 4.2 Dependence of k_H and k_D of the two-electron reduction of O_2 by AcrH$_2$ (open circles) and AcrD$_2$ (closed circles) in MeCN at 298 K on concentration of HClO$_4$.

2.0 M (Figure 4.2).[39] Such a maximal dependence of k_H and k_D on [$HClO_4$] is ascribed to the protonation of $AcrH_2$ in the presence of $HClO_4$ (eqn (10)).[39]

$$AcrH_2 + H^+ \rightleftharpoons AcrH_3^+ \qquad (10)$$

The protonation of $AcrH_2$ is indicated by the disappearance of the absorption band due to $AcrH_2$ ($\lambda_{max} = 285$ nm) in the presence of $HClO_4$.[44] In Figure 4.3, the ratios of unprotonated $AcrH_2$ and $AcrD_2$ determined by the electronic spectra in the presence of $HClO_4$ relative to the initial amount in the absence of $HClO_4$ are plotted against the $HClO_4$ concentration.[39] As shown in Figure 4.3, protonation of $AcrH_2$ or $AcrD_2$ hardly occurred in the region [$HClO_4$] < 1.0 M, whereas the fraction of unprotonated $AcrH_2$ or $AcrD_2$ decreased significantly with an increase in the $HClO_4$ concentration > 2.0 M. Thus, the decrease in k_H and k_D in the high $HClO_4$ concentration range (> 2.0 M) is ascribed to the protonation of $AcrH_2$ to afford $AcrH_3^+$, which is inactive for the reduction of O_2. The small k_H/k_D value suggests that PCET may be the rate-determining step, followed by fast proton transfer (PT) and electron transfer (ET) for the overall hydride transfer from $AcrH_2$ to O_2 in competition with protonation of $AcrH_2$ (Scheme 4.2).[39]

A solid acid is also effective for PCET from Me_2Fc and $AcrH_2$ to O_2. When a CH_2Cl_2 solution of Me_2Fc was introduced onto alumina (0.10 g), activated by calcination at 400 °C for 5 h, the formation of Me_2Fc^+ and $AcrH^+$ was observed in PCET from Me_2Fc and $AcrH_2$ to O_2 in the presence of activated alumina.[39] The ratio of product yields to alumina (mol g^{-1}) as well as the acidity of the activated alumina (mol g^{-1}) titrated by the Hammett indicator methyl red

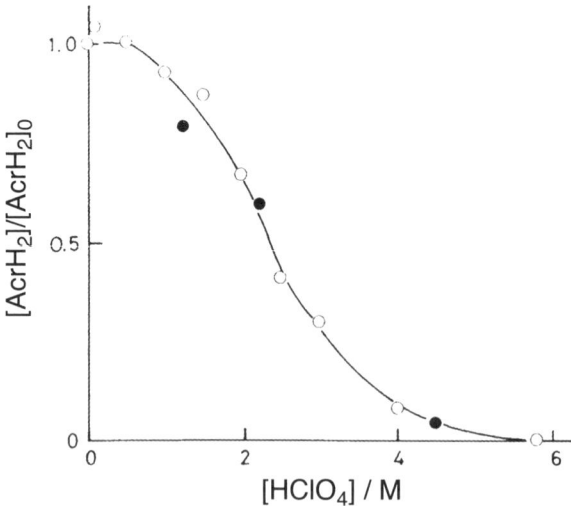

Figure 4.3 Dependence of the ratio of the concentration of free (unprotonated) $AcrH_2$ (open circles) or $AcrD_2$ (closed circles) in the presence of $HClO_4$ to the initial concentration in the absence of $HClO_4$ [$AcrH_2$ or $AcrD_2$]/[$AcrH_2$ or $AcrD_2$]$_0$ on the $HClO_4$ concentration in an aqueous solution at 298 K.

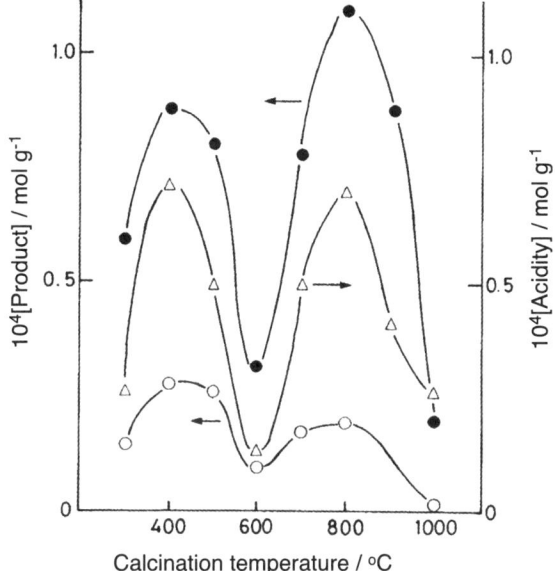

Scheme 4.2 PCET from AcrH$_2$ to O$_2$ followed by PT and ET in competition with protonation of AcrH$_2$.

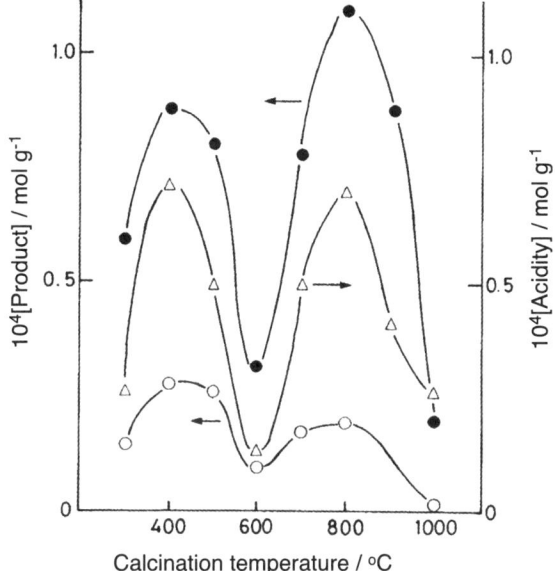

Figure 4.4 Plots of the amounts of products [AcrH$^+$ (open circles) and Me$_2$Fc$^+$ (closed circles)] formed in the heterogeneous oxidation of AcrH$_2$ and Me$_2$Fc, respectively, by O$_2$ in the presence of alumina in CH$_2$Cl$_2$ (mol g^{-1}), and the acidity ($H_0 < 4.8$) of alumina (open triangles) *versus* the calcination temperature of alumina (5 h in air).

($pK_a = 4.8$) is dependent upon the calcination temperature of alumina, as shown in Figure 4.4, where the product yields and the acidity exhibit the same maxima at 400 and 800 °C.[39] Thus, the PCET activity is determined by the acidity of the solid acid. A similar dependence of the acidity of alumina on the calcination temperature has been reported previously.[45] The product yield of AcrH$^+$ in the two-electron reduction of O$_2$ by AcrH$_2$ increased with increasing the acid strength of the solid acid in the order Al$_2$O$_3$ < SiO$_2$–Al$_2$O$_3$ ≈ HY type zeolite which was converted from NH$_4$Y-type by calcination.[46]

Cobalt(II) tetraphenylporphyrin (CoTPP) can also act as an electron donor in PCET to O$_2$.[43] In the absence of HClO$_4$, CoTPP is stable toward O$_2$ in

MeCN.[43] Upon addition of $HClO_4$ to an O_2-saturated MeCN solution of CoTPP, however, CoTPP was instantly oxidized to yield $CoTPP^+$ and H_2O_2 *via* PCET from CoTPP to O_2 (eqn (11)), followed by fast PCET from CoTPP to $HO_2^{\bullet}$ (eqn (12)).[43]

$$CoTPP + O_2 + H^+ \xrightarrow{k_{et}} CoTPP^+ + HO_2^{\bullet} \tag{11}$$

$$CoTPP + HO_2^{\bullet} + H^+ \xrightarrow{fast} CoTPP^+ + H_2O_2 \tag{12}$$

Rates of the oxidation of CoTPP by an excess amount of O_2 in the presence of $HClO_4$ in MeCN obeyed pseudo-first-order kinetics, and the pseudo-first-order rate constant increased linearly with increasing concentration of O_2.[43] Thus, the rate of formation of $CoTPP^+$ is given by eqn (13).

$$d[CoTPP^+]/dt = k_{et}[CoTPP][O_2] \tag{13}$$

The second-order rate constant (k_{et}) of PCET increased linearly with increasing concentration of $HClO_4$ as the case of PCET from Me_2Fc to O_2 (eqn (6)).[43]

The facile oxidation of CoTPP by O_2 in the presence of $HClO_4$ in MeCN has also been confirmed by a voltammetric study, as shown in Figure 4.5.[43]

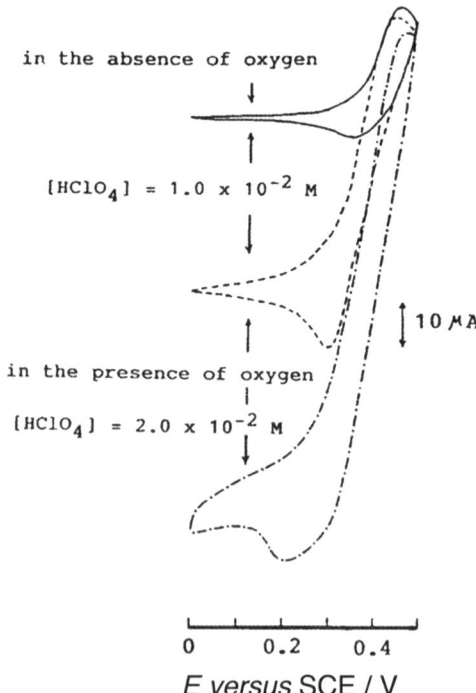

Figure 4.5 Cyclic voltammograms of $CoTPP^+$ (2.0×10^{-3} M) in the absence and presence of O_2 in MeCN containing different concentrations of $HClO_4$.

The cyclic voltammogram (CV) of $CoTPP^+$ in the absence of O_2 shows the quasi-reversible redox couple of $CoTPP^+/CoTPP$ at 0.35 V *versus* SCE, which was unaffected by the presence of $HClO_4$.[43] No reduction of O_2 has been observed in the absence of $CoTPP^+$ under otherwise the same conditions.[43] In the presence of O_2, however, the catalytic reduction of O_2 is observed at the reduction potential of $CoTPP^+$, and the reduction current increases with an increase in the $HClO_4$ concentration (Figure 4.5).[43]

4.3   MCET from Electron Donors to O_2

MCET from one-electron reductants to oxygen are related to the Lewis acidity of metal ions.[19] Charges and ion radii are important factors to determine the Lewis acidity of metal ions.[47] The binding of metal ions to $O_2^{•-}$ is clearly indicated by EPR spectra of the superoxide–metal ion complexes ($O_2^{•-}-M^{n+}$), as shown in Figure 4.6, where the superhyperfine structures due to bound metal ions ($I = 7/2$ for Lu^{3+} and Sc^{3+}) are observed as eight lines at the g_{zz} components.[48] The binding energies of a variety of metal ions with superoxide ion ($O_2^{•-}$) have been derived from the g_{zz} values of the EPR spectra of $O_2^{•-}-M^{n+}$, providing a quantitative measure of the Lewis acidity of the metal ions.[48] The EPR spectrum of $O_2^{•-}$ shows a typical anisotropic signal with $g_{//} = 2.090$ and $g_\perp = 2.005$.[49] The g_{zz}-values of $O_2^{•-}-M^{n+}$ complexes produced in the presence of a variety of closed shell metal ions become significantly smaller than the values for $O_2^{•-}$ due to the binding of metal ions to $O_2^{•-}$.[48] The deviation of the g_{zz}-value from the free spin value ($g_e = 2.0023$) is caused by the spin-orbit interaction as given by eqn (14), [50,51]

$$\Delta E = (g_{zz} - g_e)/2\lambda \tag{14}$$

where λ is the spin-orbit coupling constant (0.014 eV)[52] and ΔE is the energy splitting of the π_g levels due to the binding of M^{n+} to $O_2^{•-}$.

The ΔE value obtained from the deviation of the g_{zz} value from the free spin value increases in order: monovalent cations (M^+) < divalent cations (M^{2+}) < trivalent cations (M^{3+}).[48] The ΔE value also increases with decreasing ion radius when the oxidation state of the metal ion is the same. The same trend has been reported for $O_2^{•-}$ adsorbed on the surface of various metal oxides, which act as Lewis acids as well.[53,54] The scandium ion, which has the smallest ion radius among the trivalent metal cations, gives the largest ΔE value.[48]

MCET from CoTPP to O_2 was examined in the presence of a series of metal ions (M^{n+}, $n = 1-3$) by the UV-vis spectral change for the decay of CoTPP ($\lambda_{max} = 411$ nm) and the formation of $CoTPP^+$ ($\lambda_{max} = 434$ nm) in MeCN at 298 K, as shown in Figure 4.7.[48] As mentioned above, no electron transfer from CoTPP ($E_{ox} = 0.35$ V *versus* SCE in MeCN)[43] to O_2 ($E_{red} = -0.86$ V *versus* SCE in MeCN)[21] occurs in MeCN at 298 K.[48] In the presence of M^{n+}, however, an efficient electron transfer from CoTPP to O_2 occurs to yield $CoTPP^+$ and $O_2^{•-}-M^{n+}$ (eqn (15)).[48]

$$CoTPP + O_2 + M^{n+} \xrightarrow{k_{et}} CoTPP^+ + O_2^{•-} - M^{n+} \tag{15}$$

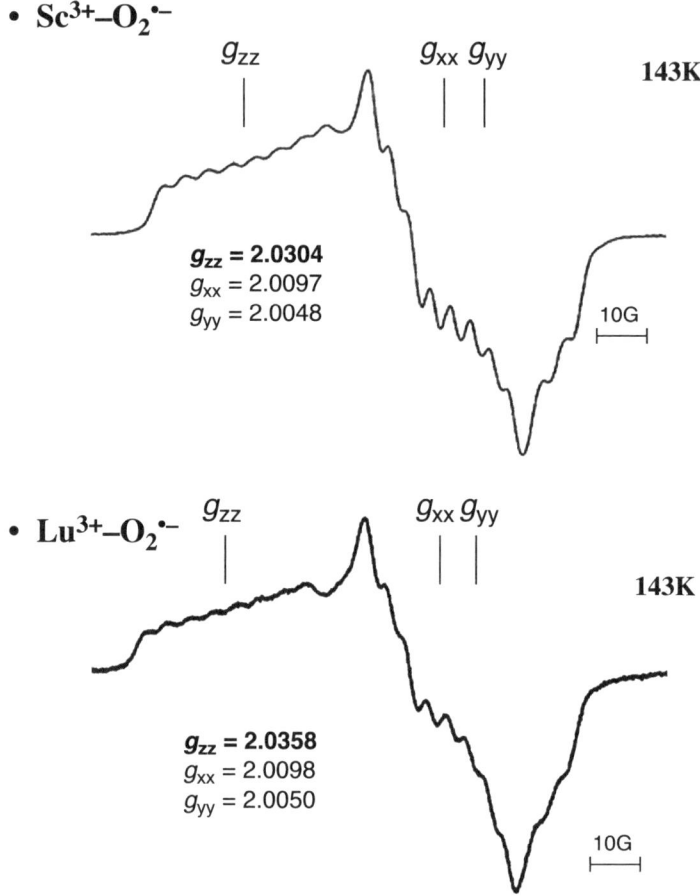

Figure 4.6 EPR spectra of $O_2{}^{\bullet-}-M^{n+}$ complexes between superoxide ion and metal ions.

The MCET rates obeyed second-order kinetics, showing a first-order dependence on the concentration of each reactant, O_2 and CoTPP.[48] The observed second-order rate constant (k_{et}) increases linearly with increasing metal ion concentration as shown in Figure 4.8, where Lu^{3+} and Mg^{2+} were used as metal ions.[48] Such a linear correlation between k_{et} and concentrations of M^{n+} indicates that electron transfer from CoTPP to O_2 is coupled with the binding of M^{n+} to $O_2{}^{\bullet-}$ (MCET) when MCET occurs in a concerted manner rather than a stepwise manner.[48] If the rate-determining step were an uphill electron transfer from CoTPP to O_2, followed by rapid binding of M^{n+} to $O_2{}^{\bullet-}$ (a stepwise pathway), the MCET rate would be independent of metal ion concentration.

The MCET rate constants (k_{MCET}) were determined from the slopes of the linear plots of k_{et} *versus* $[M^{n+}]$. There is a striking linear correlation between log k_{MCET} and the ΔE values of $O_2{}^{\bullet-}-M^{n+}$ derived from the g_{zz} values, as shown in Figure 4.9.[48] The remarkable correlation spans a range of more than 10^6 in the

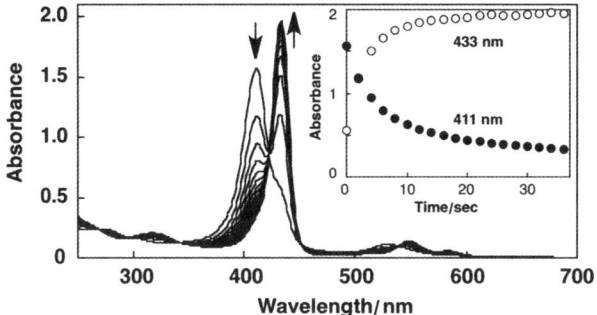

Figure 4.7 UV-vis absorption spectra observed in Sc^{3+}-coupled electron transfer from CoTPP (1.0×10^{-5} M) to O_2 (air saturated, 2.6×10^{-3} M) in the presence of scandium triflate ($Sc(OTf)_3$ 1.7×10^{-5} M) in MeCN at 298 K.

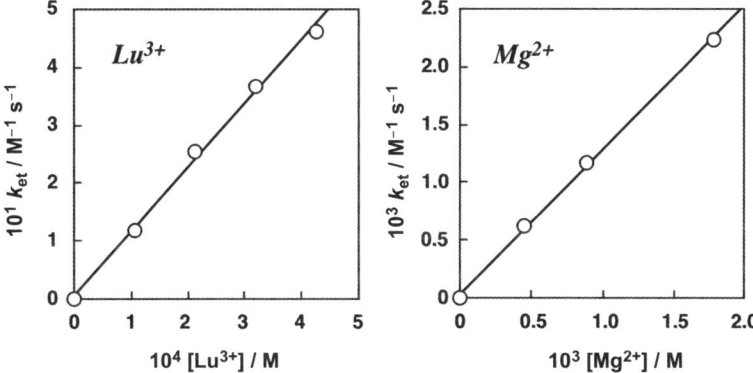

Figure 4.8 Plots of k_{et} *versus* [Lu^{3+}] and [Mg^{2+}] in MCET from CoTPP to O_2 in MeCN at 298 K.

rate constant. This correlation has been expanded to other Lewis acids including other metal ions with a variety of ligands.[55] The slope of the linear correlation between log k_{MCET} and ΔE was determined to be 14.0, which is close to the value of $1/2.3k_B T$ ($= 16.9$, where k_B is the Boltzmann constant and $T = 298$ K).[55] This indicates that the variation of ΔE is directly reflected in the difference in the activation free energy for MCET from CoTPP to O_2. Thus, the stronger the binding of M^{n+} with $O_2^{\bullet-}$, the faster the MCET rate becomes, as predicted by Scheme 4.1.

4.4 MCET from $O_2^{\bullet-}-M^{n+}$ to *p*-Benzoquinones

Electron transfer from $O_2^{\bullet-}-M^{n+}$ to electron acceptors is also coupled with the binding of M^{n+} with the radical anions of electron acceptors.[7] MCET from $O_2^{\bullet-}-M^{n+}$ to *p*-benzoquinone (Q) can be monitored by EPR when $O_2^{\bullet-}$ binds

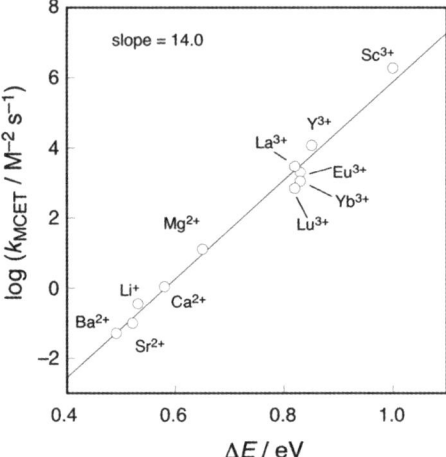

Figure 4.9   Plot of log k_{MCET} *versus* ΔE in MCET from CoTPP to O_2 in the presence of metal ions (triflate or perchlorate salts) in MeCN at 298 K.

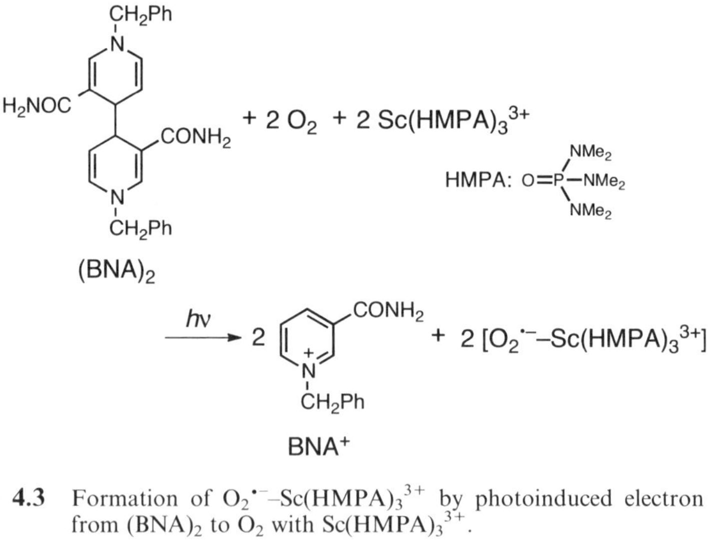

Scheme 4.3   Formation of $O_2{}^{\bullet-}$–Sc(HMPA)$_3{}^{3+}$ by photoinduced electron transfer from (BNA)$_2$ to O_2 with Sc(HMPA)$_3{}^{3+}$.

with Sc(HMPA)$_3{}^{3+}$ (HMPA = hexamethylphosphoric triamide).[56] The $O_2{}^{\bullet-}$–Sc(HMPA)$_3{}^{3+}$ complex is produced by photoinduced electron transfer from the dimeric 1-benzyl-1,4-dihydronicotinamide [(BNA)$_2$], which can act as a unique two-electron donor,[57,58] to O_2 with Sc(HMPA)$_3{}^{3+}$, as shown in Scheme 4.3. Photoinduced electron transfer from the singlet excited state of (BNA)$_2$ to O_2 occurs to produce (BNA)$_2{}^{\bullet+}$ and $O_2{}^{\bullet-}$, followed by a fast cleavage of the C–C bond of the dimer to yield the *N*-benzylnicotinamidyl radical (BNA$^{\bullet}$) and BNA$^+$.[57] The subsequent second electron transfer from BNA$^{\bullet}$ to O_2 occurs

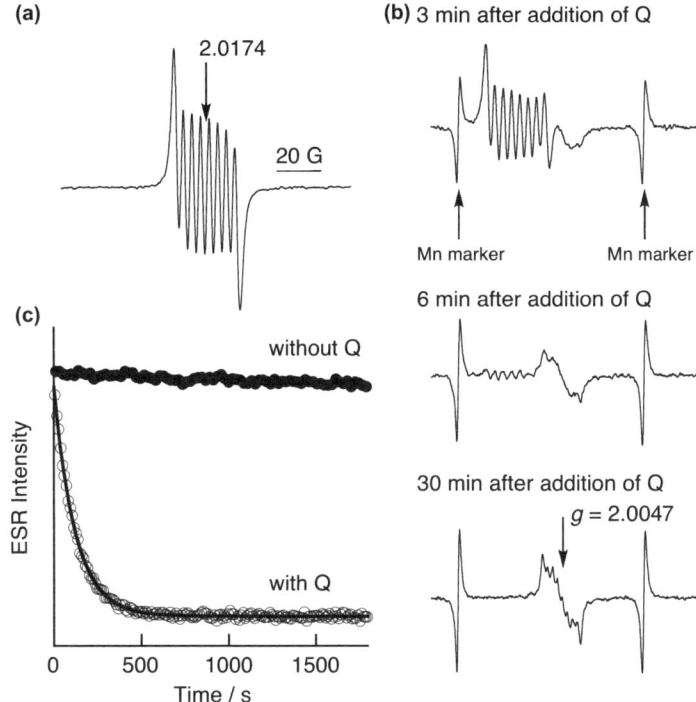

Figure 4.10 (a) EPR spectrum of $O_2{}^{\bullet-}$–Sc(HMPA)$_3{}^{3+}$ complex produced by photo-irradiation of an O_2-saturated propionitrile (EtCN) solution containing (BNA)$_2$ (2.5×10^{-2} M) and Sc(HMPA)$_3{}^{3+}$ (0.15 M). (b) Time courses of EPR signal intensity of $O_2{}^{\bullet-}$–Sc(HMPA)$_3{}^{3+}$ complex in the absence and presence of Q in EtCN at 298 K. (c) Time course of EPR spectral change in electron transfer from $O_2{}^{\bullet-}$–Sc(HMPA)$_3{}^{3+}$ to Q (5.0×10^{-2} M) in the presence of Sc(HMPA)$_3{}^{3+}$ (0.15 M) in EtCN at 298 K. The solid line is fitted by single-exponential decay.

rapidly, because the oxidation potential of BNA$^{\bullet}$ (E_{ox} *versus* SCE $= -1.08$ V)[59] is lower than the reduction potential of O_2 ($E_{red} = -0.86$ V)[38] in MeCN. Thus, two equivalents of $O_2{}^{\bullet-}$–Sc(HMPA)$_3{}^{3+}$ are produced in one photoinduced electron transfer from (BNA)$_2$ to O_2 with Sc(HMPA)$_3{}^{3+}$.[56]

The EPR spectrum of $O_2{}^{\bullet-}$–Sc(HMPA)$_3{}^{3+}$ (Figure 4.10a) exhibits a clear eight-line isotropic spectrum due to the superhyperfine coupling of $O_2{}^{\bullet-}$ with the 7/2 nuclear spin of the scandium nucleus (a(Sc) $= 3.82$ G).[56] An end-on coordination of Sc(HMPA)$_3{}^{3+}$ to $O_2{}^{\bullet-}$ was confirmed by using $^{17}O_2$; two inequivalent $a(^{17}O)$ values (21 and 14 G) were obtained, instead of an equivalent value that would be expected for a side-on form.[58] Addition of Q to an MeCN solution of $O_2{}^{\bullet-}$–Sc(HMPA)$_3{}^{3+}$ and Sc(HMPA)$_3{}^{3+}$ resulted in disappearance of the EPR signal due to $O_2{}^{\bullet-}$–Sc(HMPA)$_3{}^{3+}$ accompanied by appearance of the EPR signal due to Q$^{\bullet-}$–Sc(HMPA)$_3{}^{3+}$ (Figure 4.10b). This indicates that electron transfer from $O_2{}^{\bullet-}$–Sc(HMPA)$_3{}^{3+}$ to Q occurs to yield O_2 and Q$^{\bullet-}$–Sc(HMPA)$_3{}^{3+}$ (eqn (16)).[56]

$$O_2{}^{\bullet-} = Sc(HMPA)_3{}^{3+} \; + \; \text{[quinone]} \longrightarrow O_2 \; + \; \text{[semiquinone-Sc(HMPA)}_3{}^{3+}\text{]} \tag{16}$$

The rate of electron transfer obeyed pseudo-first-order kinetics (Figure 4.10c) and the pseudo-first-order rate constant increases linearly with increasing concentration of Q. The second-order rate constant of electron transfer (k_{et}) obtained from the slope of the linear plot of the pseudo-first-order rate constant *versus* concentration of Q increases linearly with increasing concentration of Sc(HMPA)$_3{}^{3+}$, as shown in Figure 4.11 (closed circles).[56] Similar results are obtained for coenzyme Q10 (open circles in Figure 4.11).[56] Such a linear correlation between k_{et} and concentration of Sc(HMPA)$_3{}^{3+}$ indicates that electron transfer from O$_2{}^{\bullet-}$–Sc(HMPA)$_3{}^{3+}$ to Q is coupled with binding of Sc(HMPA)$_3{}^{3+}$ to Q$^{\bullet-}$ (Scheme 4.4a).[56] The EPR spectrum of the 1:1 complex of Q10$^{\bullet-}$ and Sc(HMPA)$_3{}^{3+}$ is shown in Figure 4.12a, where the superhyperfine splitting due to one Sc(HMPA)$_3{}^{3+}$ is observed.[56]

In contrast to the case of Q and Q10, the k_{et} values of electron transfer from O$_2{}^{\bullet-}$–Sc(HMPA)$_3{}^{3+}$ to *p*-toluquinone (MeQ) and 2,5-dimethyl-*p*-benzoquinone (Me$_2$Q) exhibit second-order dependence on concentration of Sc(HMPA)$_3{}^{3+}$, as shown in Figure 4.13.[56] Such second-order dependence of k_{et} on concentration of Sc(HMPA)$_3{}^{3+}$ results from binding of two Sc(HMPA)$_3{}^{3+}$ complexes to MeQ$^{\bullet-}$ and Me$_2$Q$^{\bullet-}$, which is coupled with electron transfer from O$_2{}^{\bullet-}$–Sc(HMPA)$_3{}^{3+}$ to MeQ and Me$_2$Q (Scheme 4.4b, shown for the case of Me$_2$Q).[56] The EPR spectrum of the 1:2 complex between MeQ$^{\bullet-}$ and Sc(HMPA)$_3{}^{3+}$ is shown in Figure 4.12b, where superhyperfine splitting due to the two Sc(HMPA)$_3{}^{3+}$ complexes is observed.[56] The contribution of the second-order dependence of k_{et} on concentration of Sc(HMPA)$_3{}^{3+}$ can be

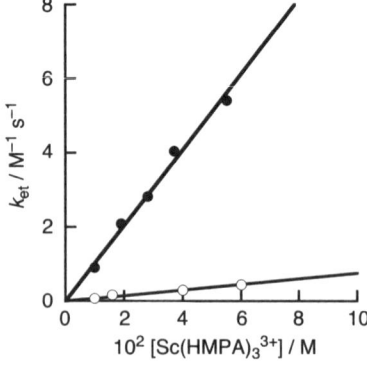

Figure 4.11   Plots of k_{et} *versus* [Sc(HMPA)$_3{}^{3+}$] for Sc(HMPA)$_3{}^{3+}$-coupled electron transfer from O$_2{}^{\bullet-}$–Sc(HMPA)$_3{}^{3+}$ complex to Q ($\bullet$), and Q10 ($\bigcirc$) in EtCN at 298 K.

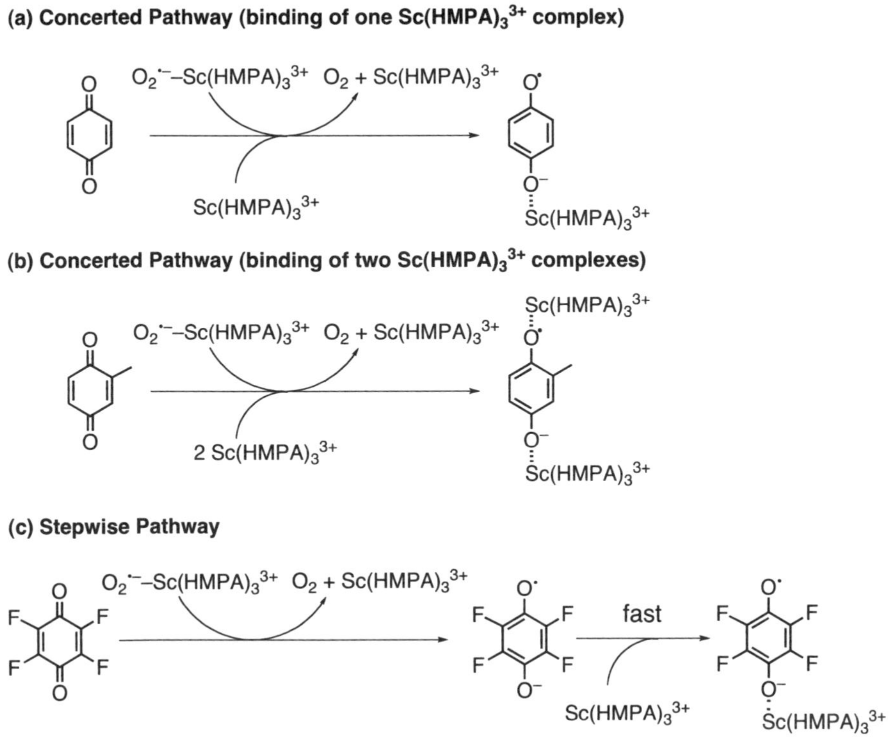

(a) Concerted Pathway (binding of one Sc(HMPA)$_3$$^{3+}$ complex)

(b) Concerted Pathway (binding of two Sc(HMPA)$_3$$^{3+}$ complexes)

(c) Stepwise Pathway

Scheme 4.4   Electron transfer from $O_2$$^{\bullet-}$–Sc(HMPA)$_3$$^{3+}$ to *p*-benzoquinone derivatives.

separated from the first-order dependence by linear plots of $k_{et}/[\text{Sc(HMPA)}_3{}^{3+}]$ *versus* [Sc(HMPA)$_3$$^{3+}$], as in Figure 4.13b.[56] In this case, a small portion of MeQ and Me$_2$Q forms 1:1 and 1:2 complexes with Sc(HMPA)$_3$$^{3+}$, and the 1:2 complex may be more reactive than the 1:1 complex.[56]

In the case of electron transfer from $O_2$$^{\bullet-}$–Sc(HMPA)$_3$$^{3+}$ to Cl$_2$Q, F$_4$Q and Cl$_4$Q, the k_{et} value increases linearly with an increase in concentration of Sc(HMPA)$_3$$^{3+}$, as shown in Figure 4.14.[56] In these cases, however, clear intercepts are observed. The intercept indicates that electron transfer from $O_2$$^{\bullet-}$–Sc(HMPA)$_3$$^{3+}$ to Cl$_2$Q, F$_4$Q, and Cl$_4$Q occurs first, followed by rapid binding of Sc(HMPA)$_3$$^{3+}$ to Cl$_2$Q$^{\bullet-}$, F$_4$Q$^{\bullet-}$, and Cl$_4$Q$^{\bullet-}$.[56] The concerted pathways *versus* the stepwise pathway in the case of Q, Me$_2$Q and F$_4$Q$^{\bullet-}$ are shown in Scheme 4.4a, 4.4b and 4.4c, respectively.[56]

4.5   Catalytic Two-Electron Reduction of O$_2$ *via* MCET and PCET

MCET from Me$_2$Fc to O$_2$ in the presence of HClO$_4$ in benzonitrile (PhCN) or MeCN at 298 K is slow, as described above (eqn (6) and (7)),[39] whereas MCET from CoTPP to O$_2$ occurs rapidly (eqn (11) and (12)).[43] These can be combined

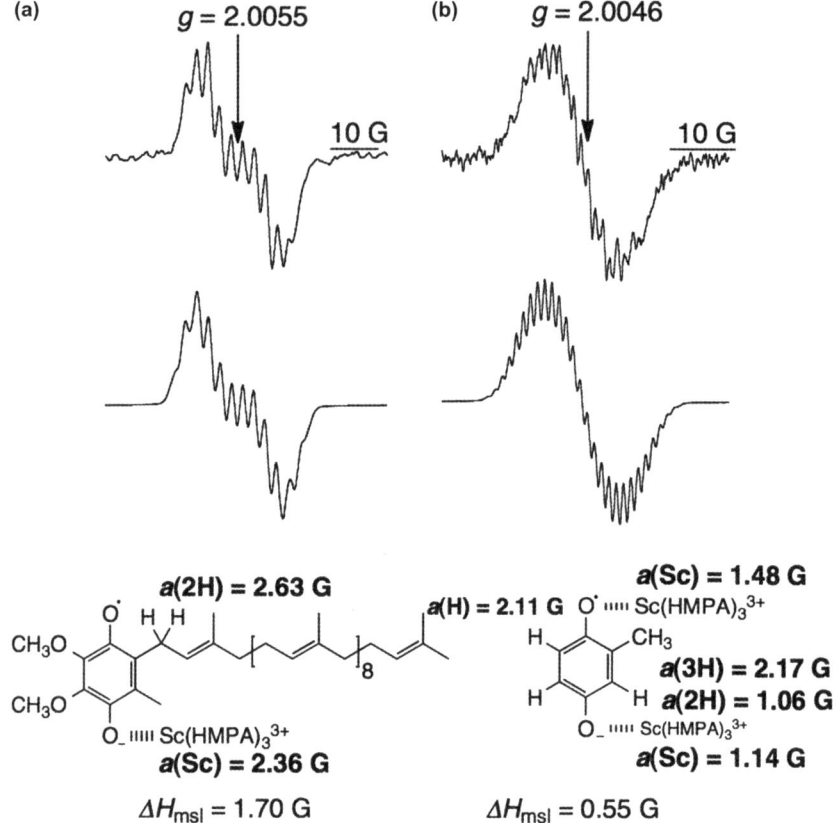

Figure 4.12 EPR spectra of EtCN solutions containing $(BNA)_2$ $(1.0 \times 10^{-3}$ M), and $Sc(HMPA)_3{}^{3+}$ $(1.5 \times 10^{-1}$ M) with (a) Q10 $(9.5 \times 10^{-3}$ M) and (b) MeQ $(1.3 \times 10^{-2}$ M), measured under photoirradiation with a high-pressure mercury lamp at 298 K. The computer simulated spectra with the corresponding g, *hfc* and ΔH_{msl} values are shown below the observed spectra.

as $CoTPP^+$-catalyzed reduction of O_2 by Me_2Fc.[43] Thus, the addition of $CoTPP^+$ to an air-saturated PhCN or MeCN solution of Me_2Fc with $HClO_4$ results in efficient PCET from Me_2Fc to O_2.[43] Upon the mixing of Me_2Fc and $CoTPP^+$ in the presence of $HClO_4$ in air-saturated MeCN at 298 K, the rise of an absorption band at 690 nm due to Me_2Fc^+ was immediately observed within 0.1 s, followed by an additional slower increase in the absorbance with a much longer reaction time (Figure 4.15).[43] The first step in Figure 4.15 corresponds to the initial two-electron reduction of O_2 with Me_2Fc, catalyzed by $CoTPP^+$ in the presence of $HClO_4$ in MeCN to yield Me_2Fc^+ and H_2O_2.[43] The slower second step in Figure 4.15 is ascribed to the further two-electron reduction of H_2O_2 by Me_2Fc to yield Me_2Fc^+ and H_2O.[43] In consequence, the oxidation of Me_2Fc with an excess amount of O_2, catalyzed by $CoTPP^+$ in the presence of $HClO_4$ results in only the two-electron reduction of O_2 to H_2O_2.[43]

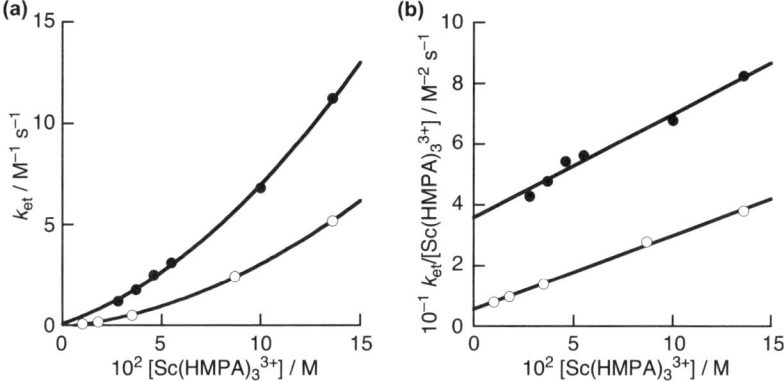

Figure 4.13 (a) Plots of k_{et} *versus* $[Sc(HMPA)_3{}^{3+}]$ and (b) $k_{et}/[Sc(HMPA)_3{}^{3+}]$ *versus* $[Sc(HMPA)_3{}^{3+}]$ for $Sc(HMPA)_3{}^{3+}$-coupled electron transfer from $O_2{}^{\bullet-}$–$Sc(HMPA)_3{}^{3+}$ complex to MeQ (●) and Me_2Q (○) in EtCN at 298 K.

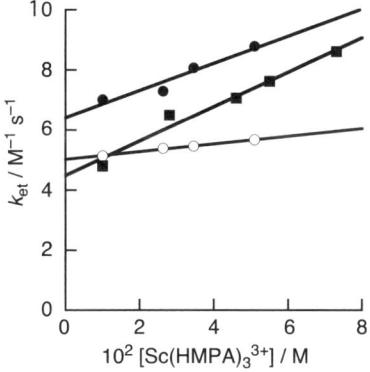

Figure 4.14 Plots of k_{et} *versus* $[Sc(HMPA)_3{}^{3+}]$ for $Sc(HMPA)_3{}^{3+}$-coupled electron transfer from $O_2{}^{\bullet-}$–$Sc(HMPA)_3{}^{3+}$ complex to Cl_4Q (●), Cl_2Q (■) and F_4Q (○) in EtCN at 298 K.

CoOEP (OEP = octaethylporphyrin dianion) also acts as an efficient catalyst for the two-electron reduction of O_2 by ferrocene derivatives in the presence of $HClO_4$ in O_2-saturated PhCN. It was confirmed that H_2O_2 was formed in the two-electron reduction of O_2 by iodometric measurements.[60] The rate of formation of the ferrocenium cation (Fc^+) in the CoOEP-catalyzed two-electron reduction of O_2 by ferrocene (Fc) in the presence of $HClO_4$ in O_2-saturated PhCN at 298 K obeyed pseudo-first-order kinetics.[60] The pseudo-first-order rate constant (k_{obs}) increases linearly with an increase in the catalyst concentration (Figure 4.16a).[60] The second-order catalytic rate constant (k_{cat}) remains constant with the change in O_2 and $HClO_4$ concentrations (Figures 4.16b and 4.16c).[60]

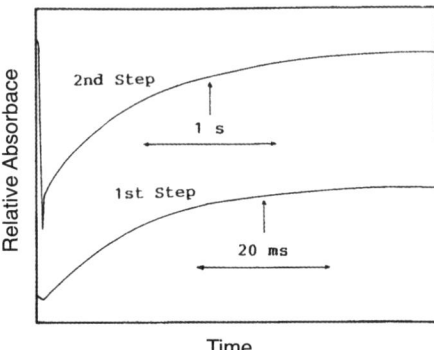

Figure 4.15 Kinetic curves for two-step oxidation of Me_2Fc, $(2.0 \times 10^{-2}$ M) by O_2
$(2.6 \times 10^{-3}$ M), catalyzed by $CoTPP^+$ $(2.0 \times 10^{-4}$ M) in the presence of
$HClO_4$ $(2.0 \times 10^{-2}$ M) in MeCN, followed by the increase in absorbance
at 690 nm due to Me_2Fc^+.

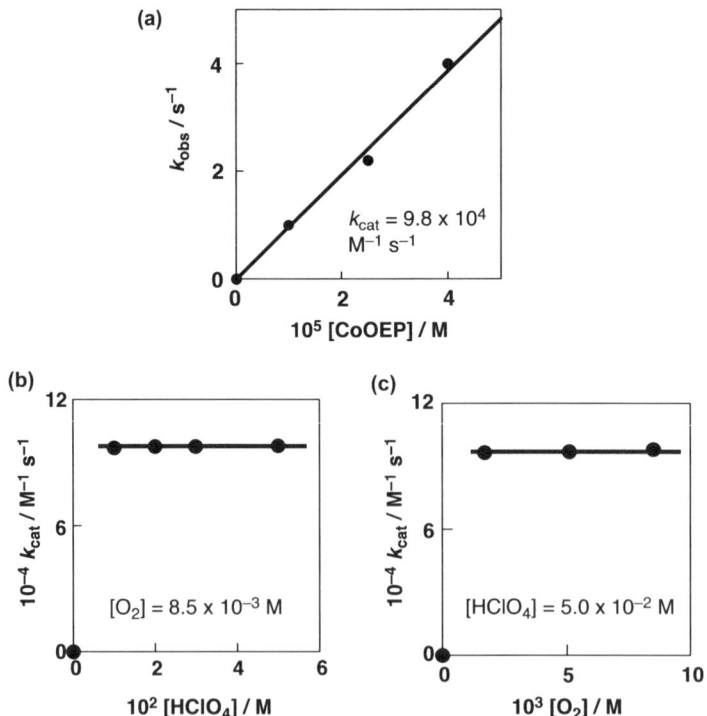

Figure 4.16 (a) Plot of k_{obs} *versus* [CoOEP] for the CoOEP-catalyzed two-electron
reduction of O_2 by Fc in the presence of $HClO_4$ in PhCN at 298 K.[60]
(b) Plot of k_{cat} *versus* $[HClO_4]$.[60] (c) Plot of k_{cat} *versus* $[O_2]$.[60]

The k_{cat} values determined from formation of Fc^+ and Me_2Fc^+ in the CoOEP-catalyzed reduction of O_2 by Fc and Me_2Fc are twice that of the k_{et} values of electron transfer from Fc and Me_2Fc to $CoOEP^+$ in the absence of O_2, respectively.[60] This indicates that the turnover-determining step (t.d.s.) for the catalytic two-electron reduction of O_2 is the electron-transfer step from ferrocene derivatives to $CoOEP^+$, as shown in Scheme 4.5.[60] In such a case, the rate of formation of Fc^+ is given by eqn (17), where the catalytic rate constant (k_{cat}) corresponds to $2k_{et}$.[60]

$$d[Fc^+]/dt = 2k_{et}[Fc][CoOEP^+] \qquad (17)$$

Electron transfer from Fc to $CoOEP^+$ occurs, followed by fast electron transfer from CoOEP to O_2, in the presence of an acid to produce the $[Co^{III}(OEP)O_2H]^+$, which is further reduced by Fc in the presence of $HClO_4$ to produce H_2O_2, accompanied by regeneration of $CoOEP^+$. The catalytic mechanism of two-electron reduction of O_2, given in Scheme 4.5, is virtually the same as that reported for CoTPP-catalyzed two-electron reduction of O_2 by ferrocene derivatives.[43]

The catalytic two-electron reduction of O_2 also occurs with cobalt corroles as well as cobalt porphyrins in PhCN.[61] The addition of [10-pentafluorophenyl-5,15-dimesityl-corrole]cobalt [$(F_5PhMes_2Cor)Co$] to an air-saturated PhCN solution of Me_2Fc and $HClO_4$ resulted in efficient reduction of O_2 by Me_2Fc. Only the two-electron reduction of O_2 occurs and there is no further reduction to produce more than two equivalents of Me_2Fc^+ (Scheme 4.6).[62] Electron transfer from Me_2Fc ($E_{ox}=0.29$ V *versus* SCE)[43] to $[(F_5PhMes_2Cor)Co]^+$ ($E_{red}=0.38$ V)[61] occurs efficiently to produce Me_2Fc^+ and $(F_5PhMes_2Cor)Co$.[62] The cobalt(III) corrole complex [$(F_5PhMes_2Cor)Co$] can reduce O_2 in the presence of $HClO_4$. The site of electron transfer was examined by EPR characterization of the singly oxidized cobalt corrole.[62]

The observed g value (2.0032) of the singly oxidized cobalt corrole, obtained by the chemical oxidation of $(F_5PhMes_2Cor)Co$ with one equivalent of

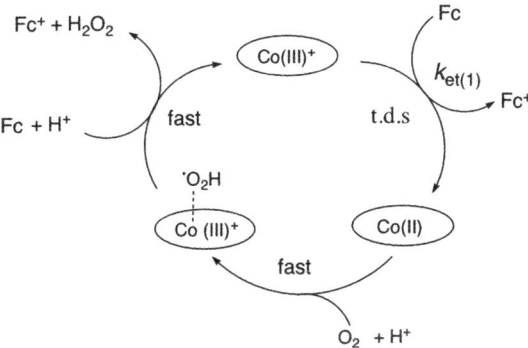

Scheme 4.5 CoOEP-catalyzed two-electron reduction of O_2 in the presence of $HClO_4$.

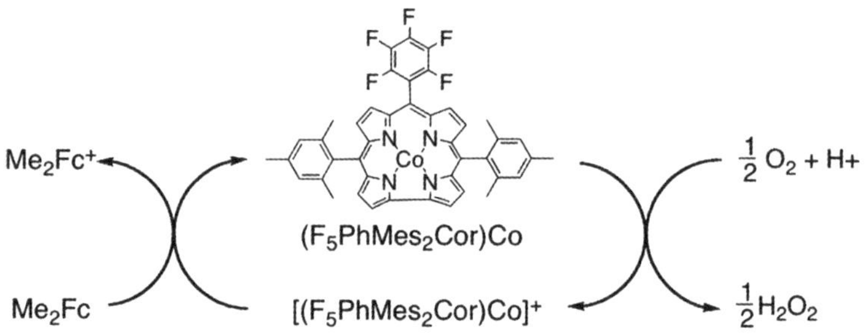

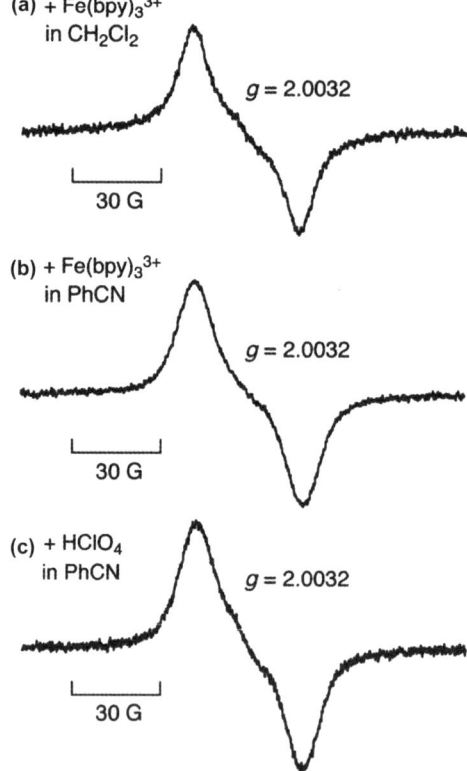

Scheme 4.6   (F$_5$PhMes$_2$Cor)Co-catalyzed two-electron reduction of O$_2$ in the presence of HClO$_4$.

(a) + Fe(bpy)$_3{}^{3+}$
in CH$_2$Cl$_2$

$g = 2.0032$

30 G

(b) + Fe(bpy)$_3{}^{3+}$
in PhCN

$g = 2.0032$

30 G

(c) + HClO$_4$
in PhCN

$g = 2.0032$

30 G

Figure 4.17   EPR spectra of singly oxidized (F$_5$PhMes$_2$Cor)Co produced by the chemical oxidation of (F$_5$PhMes$_2$Cor)Co (3.0 × 10^{-4} M) with (a) Fe(bpy)$_3$(PF$_6$)$_3$ (3.0 × 10^{-4} M) in CH$_2$Cl$_2$ at 183 K, (b) Fe(bpy)$_3$(PF$_6$)$_3$ (3.0 × 10^{-4} M) in PhCN at 298 K and (c) HClO$_4$ (2.0 × 10^{-2} M) in aerated PhCN at 298 K.

[Fe(bpy)$_3$]$^{3+}$ (bpy = 2,2′-bipyridine), is characteristic of an organic radical, as shown in Figure 4.17,[62] being quite different from the large g value (2.037) observed for cobalt(IV) porphyrin complexes.[63] Thus, the singly oxidized species is assigned as the cobalt(III) corrole radical cation rather than cobalt(IV) corrole. In contrast to the case of cobalt porphyrins (Scheme 4.5), cobalt corroles act as effective catalysts in the reduction of O_2 with $HClO_4$ *via* the redox couple between cobalt(III) corroles and cobalt(III) corrole radical cations (Scheme 4.6).[62]

4.6 Catalytic Four-Electron Reduction of O_2

4.6.1 Cofacial Dicobalt Porphyrin and Porphyrin–Corrole Dyads

When monomeric cobalt porphyrins are replaced by cofacial dicobalt porphyrin dyads and porphyrin–corrole dyads, the four-electron reduction of O_2 by ferrocene derivatives occurs efficiently, depending on the type of spacer (Sp in Figure 4.18), in the presence of $HClO_4$ in PhCN.[60,64]

When a cofacial dicobalt porphyrin $Co_2(DPX)$ is used as a catalyst instead of a single cobalt porphyrin CoOEP, the concentration of Me_2Fc^+ formed in the $Co_2(DPX)$-catalyzed reduction of O_2 by Me_2Fc is four times that of the O_2 concentration.[60] Thus, the four-electron reduction of O_2 by Me_2Fc occurs efficiently in the presence of a catalytic amount of $Co_2(DPX)$ and $HClO_4$ in PhCN. It was confirmed that no H_2O_2 was formed in the catalytic reduction of O_2 by Me_2Fc.[60]

The other cofacial dicobalt porphyrins $Co_2(DPA)$, $Co_2(DPB)$, and $Co_2(DPD)$ also catalyze the reduction of O_2 by Me_2Fc, but the amount of Me_2Fc^+ formed was less than four equivalents of O_2.[60] Thus, the clean four-electron reduction of O_2 by Me_2Fc occurs only in the case of $Co_2(DPX)$ being used as a catalyst. The same selectivity with regard to the two-electron *versus* four-electron reduction of O_2 by Me_2Fc depending on the type of spacer (Sp) was observed for the catalytic reduction of O_2 with cofacial biscobalt porphyrin-corrole complexes, shown in Figure 4.18, in the presence of $HClO_4$ in PhCN.[64]

Based on the detailed kinetic comparison of the catalytic reactivities of cofacial dicobalt porphyrins and a single cobalt porphyrin together with the detection of the reactive intermediates by EPR, the catalytic mechanism of four-electron reduction of O_2 by ferrocene derivatives is summarized in Scheme 4.7.[60] The initial electron transfer from ferrocene derivatives to the Co^{III}–Co^{III}

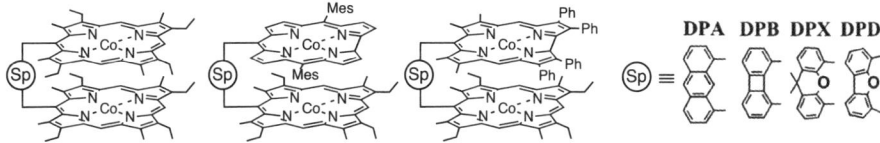

Figure 4.18 Structures of investigated biscobalt porphyrin dyads and porphyrin–corrole dyads employed for catalytic two-electron and four-electron reduction of O_2 by ferrocene derivatives.

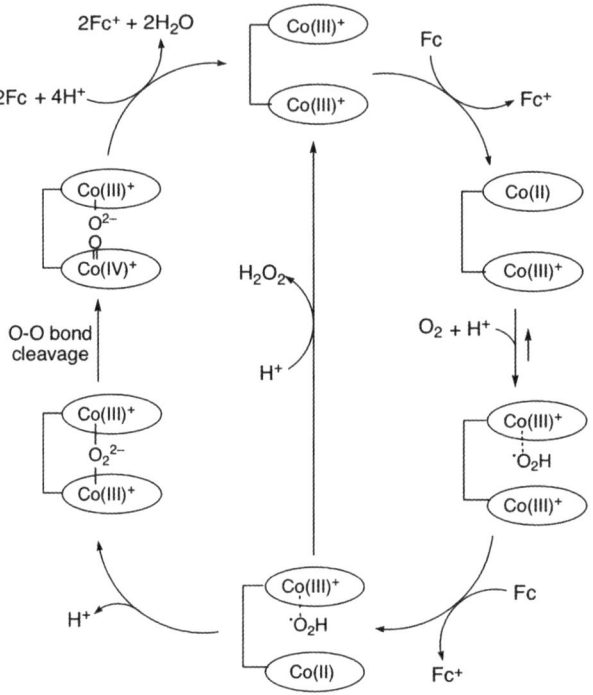

Scheme 4.7 The catalytic mechanism of four-electron reduction of O_2 by Fc with a cofacial cobalt porphyrin.[60]

complex gives the Co^{II}–Co^{III} complex, which reacts with O_2, accompanied by the reduction by ferrocene derivatives, to produce the Co^{III}–O_2–Co^{III} complex. The heterolytic O–O-bond cleavage of the Co^{III}–O_2–Co^{III} complex affords the high valent cobalt(IV)oxo porphyrin π-radical cation, which is further reduced by ferrocene derivatives in the presence of protons to yield H_2O (Scheme 4.7).[60] The critical point to distinguish between the two-electron and four-electron reduction pathways is formation of the μ-peroxo Co^{III}–O_2–Co^{III} complex, which requires an appropriate Co–Co distance in the cofacial dicobalt complex.[60] The Co–Co distance in Co_2(DPX) is best suited to the formation of the μ-peroxo Co^{III}–O_2–Co^{III} complex, resulting in the catalytic four-electron reduction of O_2.[60] In the case of monomeric cobalt porphyrins such as CoTPP and CoOEP, there is no way to form the μ-peroxo Co^{III}–O_2–Co^{III} complex, resulting in only the two-electron reduction of O_2.[60]

Thus, the interaction of two cobalt nuclei with an active form of oxygen is essential for the four-electron reduction of O_2. The μ-superoxo species of cofacial dicobalt porphyrins are produced by the reactions of cofacial dicobalt(II) porphyrins with O_2 in the presence of a bulky base (1-*tert*-butyl-5-phenylimidazole) and the subsequent one-electron oxidation of the resulting peroxo species by iodine.[60] The superhyperfine structure due to two equivalent cobalt nuclei is observed at room temperature in the EPR spectrum of the

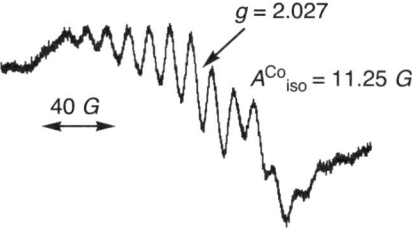

$g = 2.027$

$A^{Co}_{iso} = 11.25 \ G$

40 G

Figure 4.19 EPR spectrum of the μ-superoxo complex (*ca.* 10^{-3} M) produced by adding iodine (*ca.* 10^{-3} M) to an air-saturated PhCN solution of $Co_2(DPX)$ (5.0×10^{-3} M) in the presence of 1-*tert*-butyl-5-phenylimidazole (5.0×10^{-3} M) at 298 K.[60]

μ-superoxo species, as shown in Figure 4.19.[60] The superhyperfine coupling constant of the μ-superoxo species of $Co_2(DPX)$ determined from the computer simulation is the largest among those of cofacial dicobalt porphyrins.[60] This suggests that the efficient catalysis by $Co_2(DPX)$ of the four-electron reduction of O_2 by ferrocene derivatives results from the strong binding of the reduced oxygen with $Co_2(DPX)$, which has the most suitable distance between two cobalt nuclei for the oxygen binding.[60]

The rate of formation of Fc^+ in $Co_2(DPX)$-catalyzed four-electron reduction of O_2 by Fc in the presence of 0.05 M $HClO_4$ in O_2-saturated PhCN at 298 K also obeyed pseudo-first-order kinetics. The pseudo-first-order rate constant (k_{obs}) increases with increasing catalyst concentration of $Co_2(DPX)$, as shown in Figure 4.20a.[60] The k_{cat} values increase linearly with increasing concentrations of $HClO_4$ and O_2 (Figure 4.20b and 4.20c, respectively).[60] Such a linear dependence of k_{cat} on $[HClO_4]$ and $[O_2]$ shows a sharp contrast with the case of the CoOEP-catalyzed two-electron reduction of O_2 by Fc in Figure 4.20, where the k_{cat} values remain constant irrespective of $HClO_4$ or O_2 concentration. This indicates that the PCET from $Co^{III}Co^{II}(DPX)^+$, which is produced in the initial electron transfer from Fc and to $Co^{III}_2(DPX)^{2+}$, to O_2 is the turnover-determining step (t.d.s.) in the catalytic four-electron reduction of O_2 (Scheme 4.7).

When Fc is replaced by a much stronger reductant, that is decamethylferrocene (Fc^*), the kinetics of formation of Fc^{*+} changes drastically from pseudo-first-order kinetics in the case of Fc to zero-order kinetics, as shown in Figure 4.21a, where the rate remains constant irrespective of concentration of Fc^*.[60] The zero-order rate constant increases linearly with increasing the catalyst concentration (Figure 4.21b), but remains constant with variation of concentrations of $HClO_4$ and O_2, as shown in Figures 4.21c and d, respectively.[60] This indicates that the turnover-determining step changes from the proton-coupled electron transfer from $Co^{III}Co^{II}(DPX)^+$ to O_2 in the case of Fc to the reaction step which has nothing to do with Fc^*, $HClO_4$ or O_2. Such a process in which no electron-transfer process is involved is most likely to be O–O bond cleavage of the Co^{III}–O_2–Co^{III} complex, as shown in Scheme 4.7. The O–O bond cleavage rate is determined to be 320 s^{-1} from the slope in Figure 4.21b.[60]

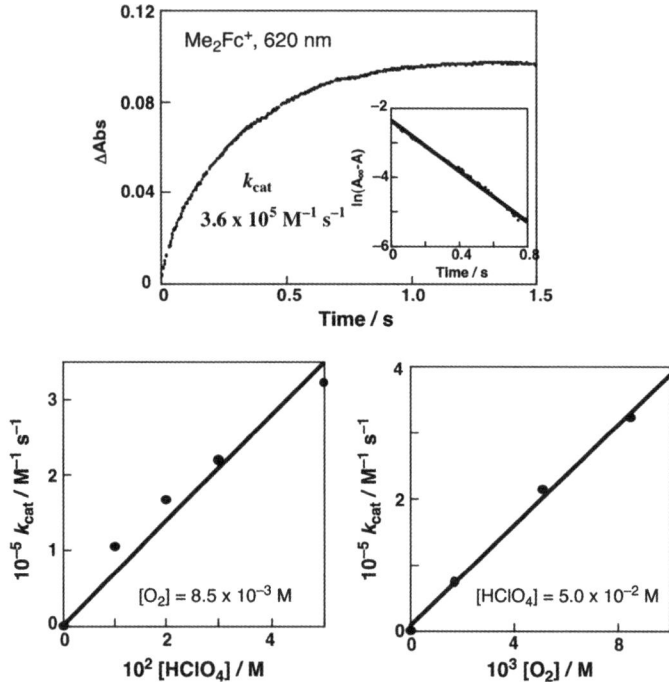

Figure 4.20   (a) Time profile of formation of Me_2Fc^+ monitored by absorbance at 620 nm ($\varepsilon = 330$ M^{-1} cm^{-1}) in electron-transfer oxidation of Me^2Fc (3.0×10^{-4} M) by catalytic dioxygen (8.5×10^{-3} M) reduction, catalyzed by $Co_2(DPX)$ (1.0×10^{-5} M) in the presence of 0.05 M $HClO_4$ in PhCN at 298 K. Inset: first-order plots. (b) Plot of k_{cat} *versus* $[Co_2(DPX)]$. (c) Plot of k_{cat} *versus* $[HClO_4]$ in the presence of 8.5×10^{-3} M O_2. (d) Plot of k_{cat} *versus* $[O_2]$ in the presence of 0.05 M $HClO_4$.[60]

When one-electron reductants (ferrocene derivatives) are replaced by two-electron reductants such as NADH analogs, 9-alkyl-10-methyl-9,10-dihydroacridines (AcrHR: R = H, Me, Et, and CH_2COOEt), the four-electron reduction of O_2 is catalyzed by $Co_2(DPA)$ to yield the 9-alkyl-10-methylacridinium ion ($AcrR^+$) and H_2O (Scheme 4.8).[65] In the case of R = Bu^t and CMe_2COOMe, however, the catalytic reduction of O_2 by AcrHR results in oxygenation of the alkyl group of AcrHR, rather than the dehydrogenation, to yield 10-methylacridinium ion ($AcrH^+$) and ROH (Scheme 4.8).[65]

In the case of $AcrH_2$, the initial slow electron transfer from $AcrH_2$ to the Co^{III}–Co^{III} complex is followed by the C(9)–H cleavage to produce $AcrH^\bullet$ in competition with the back electron transfer from the Co^{III}–Co^{II} complex to $AcrH_2^{\bullet +}$.[65,66] The catalytic rate-determining step is deprotonation of $AcrH_2^{\bullet +}$. Thus, the Co^{III}–Co^{II} complex reacts rapidly with O_2 and H^+ to give the $Co^{III}Co^{III}$ ($^\bullet O_2H$) complex, and this is followed by rapid electron transfer from $AcrH^\bullet$ to the $Co^{III}Co^{III}$ ($^\bullet O_2H$) complex to produce $AcrH^+$ and the Co^{III}-Co^{III}($^\bullet O_2H$) complex. After deprotonation, the μ-peroxo Co^{III}–O_2–Co^{III}

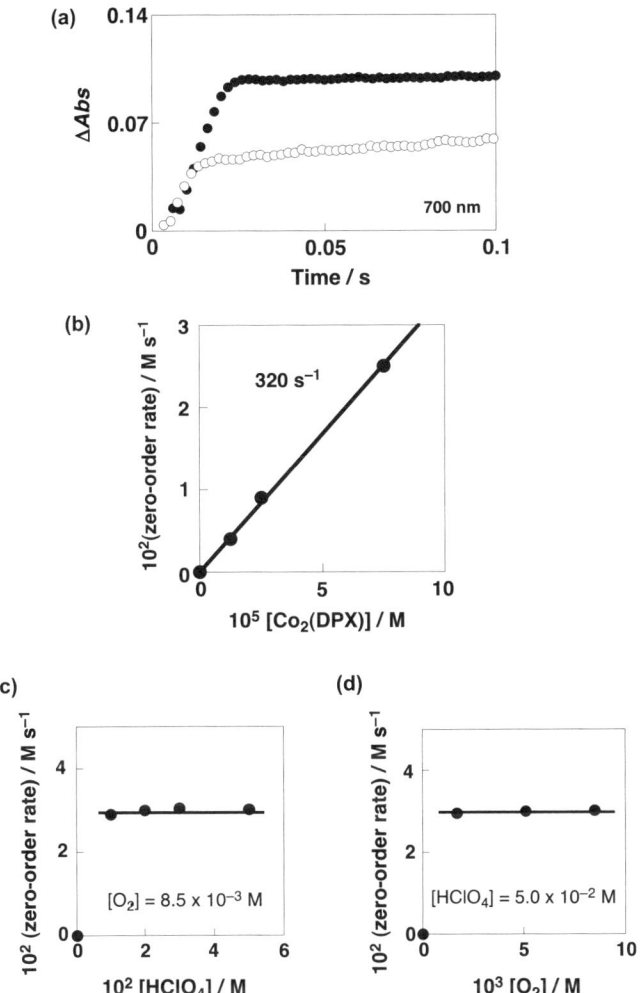

Figure 4.21 (a) Time profiles of formation of Fc^{*+} monitored by absorbance at 700 nm in the four-electron reduction of O$_2$ by Fc* [2.5 × 10^{-4} M (○), 4.0 × 10^{-4} M (●)], catalyzed by Co$_2$(DPX) (8.0 × 10^{-5} M) in the presence of 0.05 M HClO$_4$ in PhCN at 298 K.[60] (b) Plot of the zero-order rate constant *versus* [Co$_2$(DPX)].[60] (c) Plot of the zero-order rate *versus* [HClO$_4$].[60] (d) Plot of the zero-order rate *versus* [O$_2$].[60]

complex is formed, as in the case of the catalytic four-electron reduction of O$_2$ by ferrocene derivatives.[60] The heterolytic O–O-bond cleavage of the Co^{III}–O$_2$–Co^{III} complex affords the high valent cobalt(IV)oxo porphyrin π-radical cation, which is readily reduced by AcrH$_2$ in the presence of protons to yield H$_2$O, accompanied by formation of AcrH$^+$.

As in the case of the catalytic reduction of O$_2$ by ferrocene derivatives, monomeric cobalt porphyrins catalyze only the two-electron reduction of

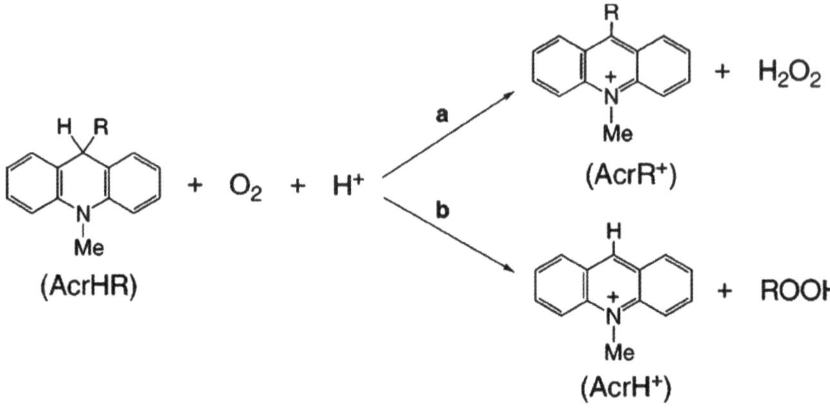

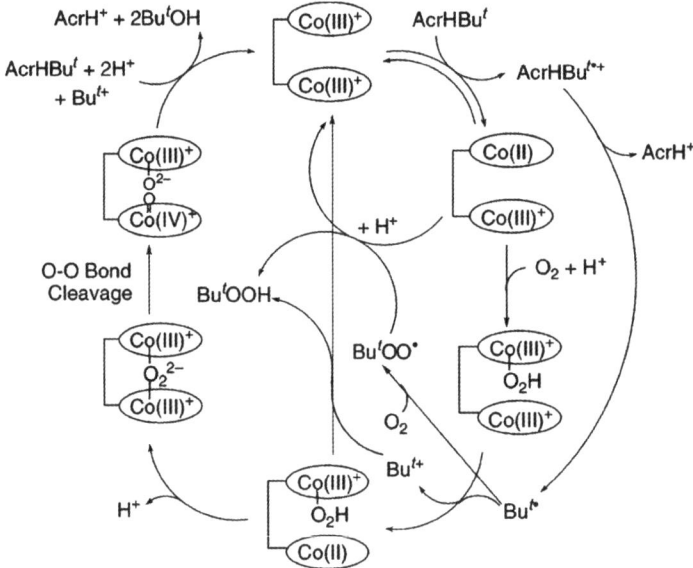

Scheme 4.8   Catalytic dehydration *versus* oxygenation of the R group of AcrHR with Co$_2$(DPA).[65]

Scheme 4.9   Mechanism of the catalytic oxygenation of the R group of AcrHR by O$_2$ with a cofacial cobalt porphyrin.[65]

oxygen by AcrH$_2$ in the presence of H$^+$.[67] In the case of AcrHBu^t, the mechanism of the catalytic four-electron reduction of O$_2$, accompanied by the oxygenation of AcrHBu^t, is modified, as shown in Scheme 4.9.[65] The initial electron transfer from AcrHBu^t to the Co^{III}–Co^{III} complex results in the homolytic C(9)–C bond cleavage to produce Bu$^{t\bullet}$ and AcrH$^+$.[68,69] Since the homolytic C(9)–C bond cleavage is also the catalytic rate-determining step, the Co^{III}Co^{III}(·O$_2$H) complex is formed by the reaction of O$_2$ and H$^+$, followed by electron transfer from $Bu^{t\bullet}$ to the $Co^{III}Co^{III}(^{\bullet}O_2H)$ complex to produce Bu^{t+} and the $Co^{III}Co^{III}(^{-}O_2H)$ complex. The subsequent step may be the same as the case of the four-electron reduction of O_2 by $AcrH_2$. The high valent cobalt-(IV)oxo porphyrin π-radical cation is formed by the heterolytic O–O-bond cleavage of the Co^{III}–O_2–Co^{III} complex, being reduced by $AcrHBu^t$ in the presence of H^+ to yield Bu^tOH, accompanied by formation of $AcrH^+$ (Scheme 4.9). However, the $Bu^{t\bullet}$ produced in the initial electron transfer from $AcrHBu^t$ to the Co^{III}–Co^{III} complex is readily trapped by O_2 to give the peroxyl radical $Bu^tOO^{\bullet}$. Such alkylperoxyl radicals ($ROO^{\bullet}$) are regarded as rather strong one-electron oxidants judging from the highly positive one-electron reduction potentials.[70] Thus, the initial electron transfer from $AcrHBu^t$ to the Co^{III}–Co^{III} complex is followed by the subsequent electron transfer from the Co^{III}–Co^{II} complex to $Bu^tOO^{\bullet}$ to produce Bu^tOOH after protonation, accompanied by regeneration of the Co^{III}–Co^{III} complex (Scheme 4.9).[65]

4.6.2 Mononuclear Cu Complexes

Multicopper oxidases, such as laccase, also activate oxygen at a site containing a three-plus-one arrangement of four Cu atoms, and are known to catalyze the four-electron reduction of O_2.[71] Electrocatalytic reduction of O_2 has been performed to probe the catalytic reactivity of synthetic CcO model complexes,[34,72] and some copper complexes have also been utilized as electrocatalysts.[73–75] In a homogeneous system, a copper complex $[(L)Cu^{II}]^{2+}$ (L = tris(2-pyridylmethyl)amine)[76] can efficiently catalyze the four-electron reduction of O_2 by one-electron reductants, such as ferrocene derivatives, in the presence of $HClO_4$ in acetone, as shown in Scheme 4.10.[77] The initial electron transfer from Fc^* to $[(L)Cu^{II}]^{2+}$ with O_2 affords the two-electron reduction of

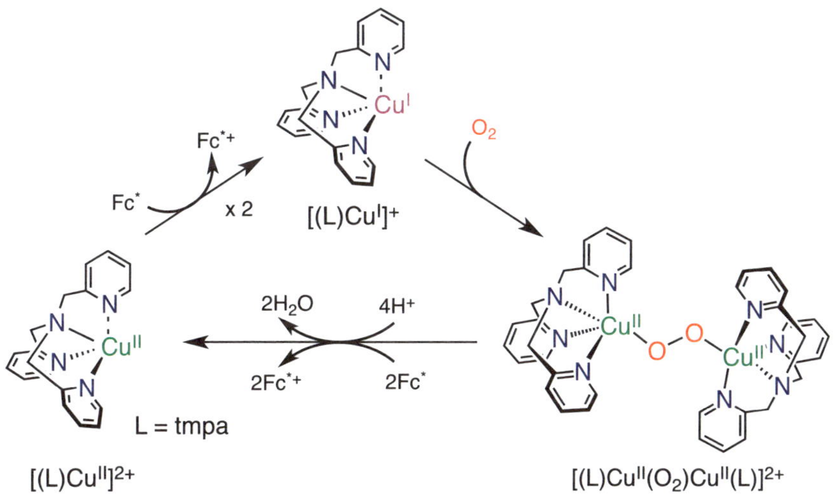

Scheme 4.10 Catalytic cycle for four-electron reduction of O_2 by Fc^* with $[(L)Cu^{II}]^{2+}$ in the presence of $HClO_4$ in acetone.[77]

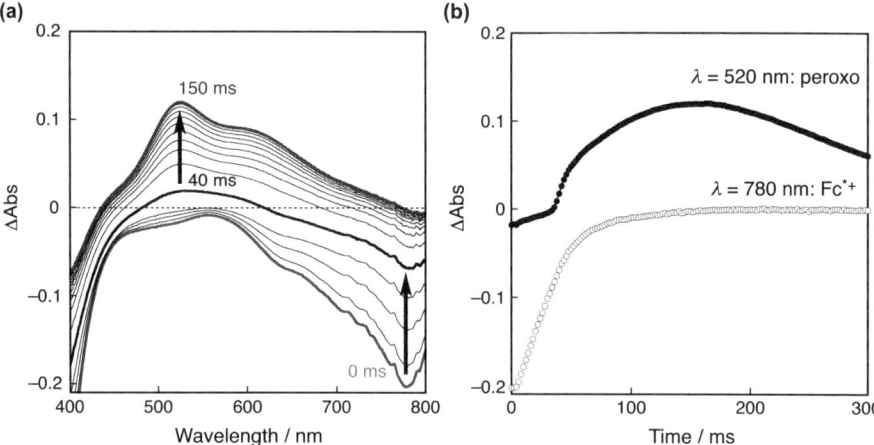

Figure 4.22 (a) Formation of the peroxo species ($\lambda_{max} = 520$ nm) in electron transfer from Fc* (1.0 mM) to [(L)Cu^{II}]$^{2+}$ (0.12 mM) in the presence of HClO$_4$ (0.35 mM) in aerated acetone at 298 K. (b) Time profile of absorbance at 520 nm (●) and 780 nm (○) due to peroxo species and Fc^{*+}, respectively.[77]

oxygen to produce the peroxo species that can be further reduced in the presence of an acid to complete the catalytic cycle for the four-electron reduction of O$_2$ by Fc* in the presence of HClO$_4$.[77]

Figure 4.22a shows the observed absorption spectral change during the catalytic reaction. Under the conditions employed, with relative concentrations of reagents as given in the caption of Figure 4.22, it is only after Fc^{*+} ($\lambda_{max} = 780$ nm) is completely formed that the peroxo species, [(L)Cu^{II}(O$_2$)-Cu^{II}(L)]$^{2+}$ ($\lambda_{max} = 520$ nm)[78] starts to be produced. {Note that the spectra recorded were taken as difference spectra in which the final spectrum was subtracted; recovery of bleaching absorption at 780 nm corresponds to the formation of Fc^{*+}.[77]} This is more clearly seen in Figure 4.22b: the time profiles for the absorbance at 780 nm due to Fc^{*+}, by comparison to the absorbance at 520 nm due to [(L)Cu^{II}(O$_2$)Cu^{II}(L)]$^{2+}$.[77] Because the concentration of HClO$_4$ is smaller than that of Fc*, HClO$_4$ has been consumed when the reaction is over.[77] It is well established that [(L)Cu^{I}]$^+$ reacts with O$_2$ affording the superoxo species which reacts rapidly with [(L)Cu^{I}]$^+$ to produce the peroxo species.[78] These processes can also be regarded as MCET, in which the Cu ion acts as both an electron donor and an Lewis acid, because the binding of the Cu^{II} ion to O$_2^{\cdot-}$ and the peroxo species promotes the electron transfer from [(L)Cu^{I}]$^+$ to O$_2$ and the superoxo species, respectively. Thus, electron-transfer reduction of [(L)Cu^{II}]$^{2+}$ by Fc* with O$_2$ but without HClO$_4$ affords [(L)Cu^{II}(O$_2$)Cu^{II}(L)]$^{2+}$.[77]

Because Cu complexes can catalyze the four-electron reduction of O$_2$ (*vide supra*), Cu-based catalysts can replace Pt-based catalysts as electrocatalyts over a wide range of pHs, albeit with moderate overpotentials, particularly in acid.[74,79] Rotating ring-disk electrode (RRDE) voltammetry results for catalysts

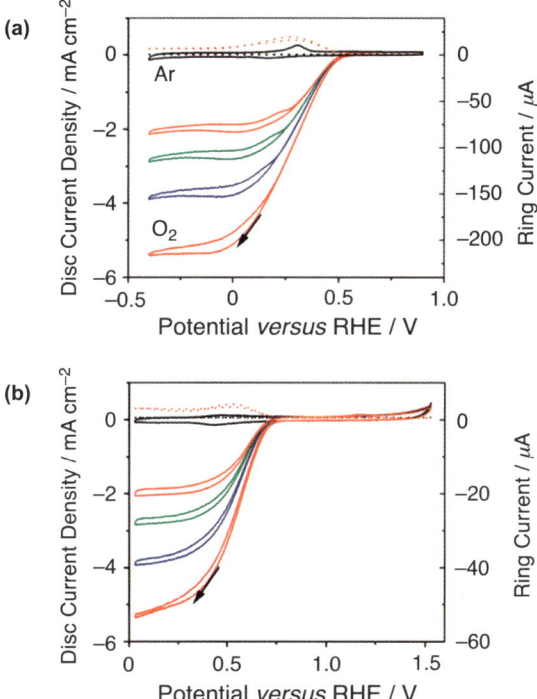

Figure 4.23 RRDE of $[(L)Cu^{II}]^{2+}$ in 0.1 M $HClO_4$ at pH 1 (a) and in pH 10 Britton–Robinson buffer (b) with ring currents (dotted lines) under 1 atm of Ar at 1600 rpm (black) and under 1 atm of O_2 at 1600 rpm (red), 800 rpm (blue), 400 rpm (green), and 200 rpm (orange).[80]

prepared from $[(L)Cu^{II}]^{2+}$ in 0.10 M $HClO_4$ are shown in Figure 4.23a.[80] At pH 1 under Ar, little current is observed, with the exception of a reversible couple at 0.23 V due to the $Cu^{I/II}$ couple.[81] With the addition of O_2, an increase in the cathodic catalytic current for the O_2 reduction is observed. The onset of the reduction current is 0.53 V *versus* RHE.[80] The ring current reveals that at potentials between the onset and the diffusion-limited current, hydrogen per-oxide is formed. The number of electrons transferred during oxygen reduction was calculated from the magnitude of the steady-state limiting current values, which were taken at a fixed potential on the catalytic wave plateaus of the dif-ferent current–voltage curves. If mass-transport alone controls the reduction of O_2 at the modified electrode, then the relationship between the limiting current and rotation rate should obey the Levich equation (eqn (18)):[82–84]

$$i_{Lev} = 0.62nFAv^{-1/6}D^{2/3}[O_2]\omega^{1/2} \tag{18}$$

where n is the number of electrons transferred, F is the Faraday constant (96485 C mol^{-1}), A is the electrode area (cm^2), n is the kinematic viscosity of the solution ($cm^2 s^{-1}$), D is the O_2 diffusion constant ($cm^2 s^{-1}$), [O_2] is the bulk concentration of O_2 (M), and ω is the rate of rotation (rad s^{-1}). The diffusion-limited current changes with the rotation rate, which, after fitting to eqn (18), shows that 3.8 e^- were transferred in the reduction.[80]

The voltammetry of [(L)Cu^{II}]$^{2+}$ is similar at pH 1 and 10 (Figure 4.23).[80] Under Ar, the $Cu^{I/II}$ couple is positively shifted from 0.23 V at pH 1 to 0.49 V at pH 10, which corresponds to a shift of 30 mV per pH unit.[80] The onset of O_2 reduction at pH 10 is at 0.77 V, which is 240 mV more positively shifted as compared to the value at pH 1.[80] The ring current also shows that H_2O_2 is formed at potentials between the onset and the diffusion-limited current.[80] The Koutecky–Levich analysis of the dependence of the rotation rate on the reduction current also shows that 3.9 e^- were transferred at this pH,[80] indicating nearly four-electron reduction of O_2. This is consistent with the results in the homogeneous system (*vide supra*).

4.6.3   A Heterodinuclear Iridium–Ruthenium Complex

A heterodinuclear iridium–ruthenium complex [Ir^{III}(Cp^*)(H_2O)-(bpm)Ru^{II}(bpy)$_2$](SO$_4$)$_2$ (Ir^{III}-OH$_2$, $Cp^* = \eta^5$-pentamethyl-cyclopentadienyl, bpm = 2,2'-bipyrimidine, bpy = 2,2'-bipyridine) can also act as an effective catalyst for removal of dissolved O_2 by the four-electron reduction of O_2 with formic acid in water at an ambient temperature.[85] The Ir^I complex reacts efficiently with O_2, as shown by the spectral titration in Figure 4.24, where the absorption spectra due to Ir^I are changed to those of Ir^{III}-OH$_2$.[85] The titration curve in the inset of Figure 4.24 indicates that Ir^I reacts with 0.5 equiv of O_2 to produce Ir^{III}-OH$_2$.

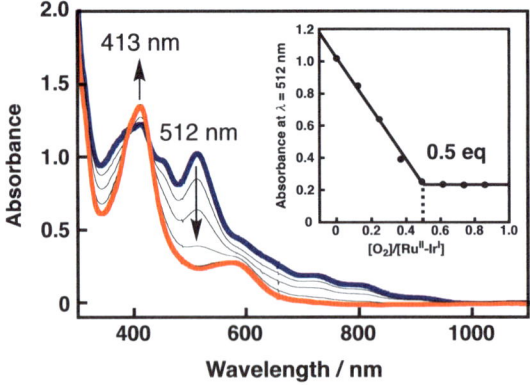

Figure 4.24   UV-vis-NIR absorption spectral changes in the reaction of Ir^I with various concentrations of O_2 in H_2O (pH 5.2) at 298 K. Ir^I is produced by the reaction of Ir^{III}-OH$_2$ (1.0 mM) with HCOOH/HCOONa (1.0 mM). Inset: Plot of absorbance at $\lambda = 512$ nm due to Ir^I *versus* the ratio of [O_2]/[Ir^I].[85]

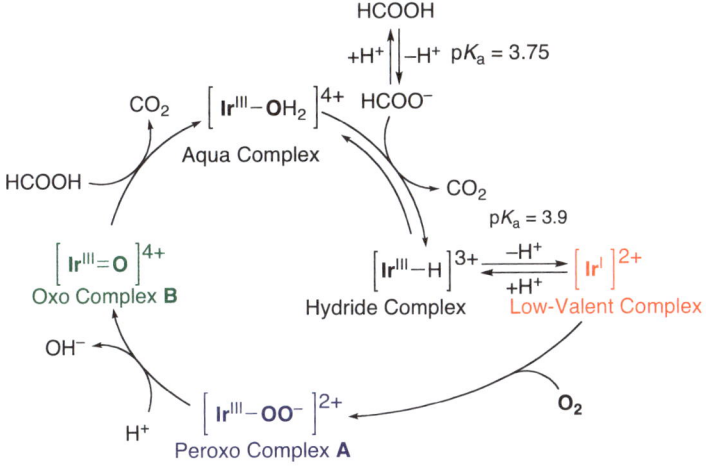

Scheme 4.11 The catalytic mechanism of the four-electron reduction of oxygen by formic acid with a heterodinuclear iridium–ruthenium complex in which the Ir center acts as the catalytically active center.[85]

Thus, the four-electron reduction of O_2 by Ir^{III}-H occurs efficiently to produce Ir^{III}-OH_2 (eqn (19)).[85] This is followed by the catalytic decomposition of formic acid to hydrogen and CO_2.[86]

$$Ir^I + 1/2\,O_2 + 2H^+ \rightarrow Ir^{III}\text{-}OH_2 \tag{19}$$

The catalytic mechanism of the four-electron reduction of oxygen with formic acid is shown in Scheme 4.11.[85] The hydride complex, which is produced by the reaction of Ir^{III}-OH_2 with $HCOO^-$, deprotonates to generate the low-valent complex Ir^I.[83] The Ir^I complex reacts with O_2 to produce an iridium(v)-oxo complex (**B**) and water *via* formation of the iridium(III)-peroxo complex (**A**).[85] The formation of an iridium-peroxo complex by the reaction of a low-valent iridium complex with O_2 has been well established.[87,88] The formation of an iridium(v)-oxo complex with cleavage of the O–O bond of an iridium(III)-peroxo complex has also been reported.[89,90] The oxo complex (**B**) reacts with HCOOH to reproduce Ir^{III}-OH_2.[85] Each intermediate in the catalytic cycle in Scheme 4.11 has been detected by stopped flow measurements.[85] Because formic acid is used as a reductant for the catalytic four-electron reduction of O_2, the water-soluble Ir catalyst can remove dissolved O_2 with formic acid completely at an ambient temperature.

4.6.4 Mononuclear Mn Complexes

As described above, metal-peroxo complexes act as intermediates in the catalytic reduction of O_2. A monomeric Mn^{II}-peroxo complex can also be prepared directly from the reaction of an Mn^{II} complex, $[Mn^{II}H_2bupa]^-$ ($[H_2bupa]^{3-}$: bis[(N'-*tert*-butylurealy)-N-ethyl]-(6-pivalamido-2-pyridylmethyl)aminato),

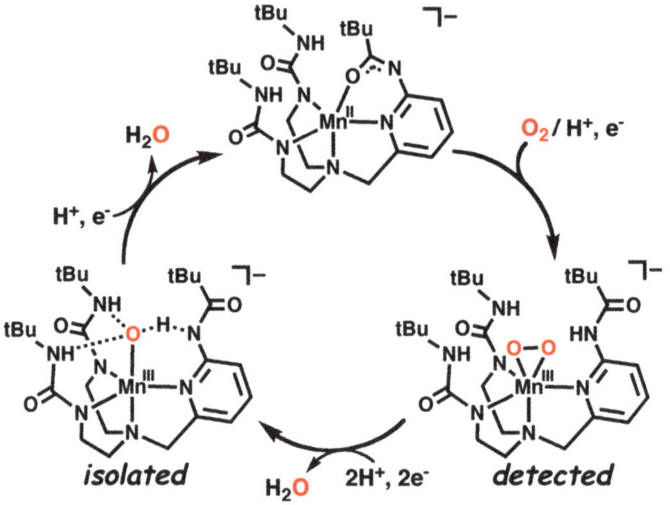

H₂O

H⁺, e⁻

O₂/ H⁺, e⁻

isolated

H₂O   2H⁺, 2e⁻   **detected**

Scheme 4.12   Proposed catalytic cycle for the four-electron reduction of O_2 to H_2O, with [MnH₂bupa]⁻ as the catalyst.[92]

with O_2.[91] The Mn^{II}-peroxo complex serves as the key intermediate in the four-electron reduction of O_2 with 1,2-diphenylhydrazine (DPH) or hydrazine as the reductant, which provides two electrons and two protons at room temperature.[92] The catalytic cycle is shown in Scheme 4.12, which highlights the role of the carboxyamido group.[92] Treating K[Mn^{II}H₂bupa] with O_2 presumably affords a superoxo species, which was not detected even at lower temperatures (less than –50 °C). It was proposed that the superoxo species was converted to the observable Mn^{III}-peroxo complex *via* hydrogen abstraction from the reductant.[92] The proton that resides on the carboxyamido group is suggested to cause its carbonyl to no longer bind to the metal center, and instead rotate in such a way as to form an intramolecular H-bond.[92] This structural change is supported by the molecular structure of [Mn^{II}H₂bupa]⁻.[92] Then PCET results in homolytic cleavage of the O–O bond in [Mn^{III}H₃bupa(O_2)]⁻, leading to formation of H_2O.[64] The formation of a $Mn^{III}(OH)_2$ complex that can deprotonate the carboxyamido group may afford an equivalent of H_2O and a Mn^{III}-O(H) complex: a hybrid Mn^{III}-O(H) complex that has strong intramolecular H-bonds.[92] Another possibility is a Mn^{IV}-oxo complex, which are known to form from O_2 activation, because the Mn^{III}-O(H) complex can be chemically oxidized to a high-spin Mn^{IV}-oxo complex that exhibits spectroscopic properties similar to those of the previously reported [Mn^{IV}H₃buea(O)]⁻ ([Hbuea]³⁻: tris(*tert*-butylureaylethlene)-aminato).[93] However, the Mn^{IV}-oxo complex was not detected during the catalytic conversion of O_2 to H_2O.[92] Homolytic cleavage of a N–H bond in DPH by the Mn^{III}-O(H) complex produces the second equivalent of H_2O. The release of H_2O from the complex may be assisted by the rebinding of the carboxyamido group to afford the starting Mn^{II} complex.[92] In the catalytic cycle (Scheme 4.12), for every equivalent of O_2 reduced, four equivalents of hydrogen atoms (*i.e.*, 2 equiv of DPH) are consumed and two equivalents of H_2O and azobenzene are produced.[92] Thus, $[Mn^{II}H_2bupa]^-$ acts as a catalyst for the four-electron reduction of O_2 to H_2O at room temperature.[92]

4.7 Summary and Conclusions

As described above, PCET and MCET play essential roles in the electron-transfer reduction of O_2 by protonation of $O_2^{\cdot-}$ and binding of metal ions to $O_2^{\cdot-}$, respectively. In both cases, formation of a strong H–O bond in $HO_2^{\cdot}$ and an M^{n+}–O bond in $O_2^{\cdot-}$–M^{n+} makes the electron-transfer reduction of O_2 energetically possible, which would otherwise be highly endergonic. The PCET and MCET reactivity is determined by the acidity of Brønsted acids and Lewis acids. PCET and MCET also play essential roles in the further reduction of $HO_2^{\cdot}$ and $O_2^{\cdot-}$–M^{n+}, leading to the two-electron and four-electron reduction of O_2. In particular MCET from redox active metal complexes, such as Cu and Mn complexes, to O_2 involves the binding of the oxidized metal complexes with the reduced oxygen species to yield the superoxo and peroxo complexes. The cleavage of the O–O bond in metal-peroxo complexes is required for the four-electron reduction of O_2 associated with the PCET of metal-peroxo complexes. Such a combination of MCET and PCET will provide valuable insights into controlling catalytic two-electron *versus* four-electron reduction of O_2, expanding the scope of both MCET and PCET processes.

Acknowledgements

The author gratefully acknowledges the contributions of his collaborators and coworkers mentioned in the references. The authors acknowledge continuous support of their study by Grants in-Aid from the Ministry of Education, Culture, Sports, Science and Technology, Japan and KOSEF/MEST through WCU project (R31-2008-000-10010-0) in Korea.

References

1. D. T. Sawyer, *Oxygen Chemistry*, Oxford University Press, New York, 1991.
2. J. P. Klinman, *J. Biol. Inorg. Chem.*, 2001, **6**, 1.
3. L. I. Simándi, in *Catalytic Activation of Dioxygen by Metal Complexes*, Kluwer, Dordrecht, The Netherlands, 1992, Chapter 1, pp. 8.
4. R. Prabhakar, P. E. M. Siegbahn, B. F. Minaev and H. Ågren, *J. Phys. Chem. B*, 2002, **106**, 3742.
5. (a) S. Fukuzumi, *Bull. Chem. Soc. Jpn.*, 1997, **70**, 1; (b) S. Fukuzumi, *Bull. Chem. Soc. Jpn.*, 2006, **79**, 177.
6. S. Fukuzumi and S. Itoh, *Antioxidants and Redox Signaling*, 2001, **3**, 807.
7. S. Fukuzumi, in *Electron Transfer in Chemistry*, ed. V. Balzani, Wiley-VCH, Weinheim, 2001, vol. 4, pp. 3.
8. D. T. Sawyer and J. S. Valentine, *Acc. Chem. Res.*, 1981, **14**, 393.

9. S. Fukuzumi and S. Itoh, in *Advances in Photochemistry*, ed. D. C. Neckers, D. H. Volman and G. von Bünau, Wiley, New York, 1998, vol. 25, pp. 107.

10. (a) S. Fukuzumi, K. Ishikawa, K. Hironaka and T. Tanaka, *J. Chem. Soc., Perkin Trans. 2*, 1987, 751; (b) S. Fukuzumi, S. Mochizuki and T. Tanaka, *J. Am. Chem. Soc.*, 1989, **111**, 1497.

11. (a) S. Fukuzumi, M. Chiba and T. Tanaka, *J. Chem. Soc., Chem. Commun.*, 1989, **14**, 941; (b) S. Fukuzumi, S. Kuroda, T. Goto, K. Ishikawa and T. Tanaka, *J. Chem. Soc., Perkin Trans.*, 1989, **2**, 1047; (c) S. Fukuzumi, M. Ishikawa and T. Tanaka, *J. Chem. Soc., Perkin Trans.*, 1989, **2**, 1811.

12. (a) S. Fukuzumi, S. Mochizuki and T. Tanaka, *J. Phys. Chem.*, 1990, **94**, 722; (b) K. Ishikawa, S. Fukuzumi, T. Goto and T. Tanaka, *J. Am. Chem. Soc.*, 1990, **112**, 1577.

13. (a) S. Fukuzumi, S. Kuroda and T. Tanaka, *J. Am. Chem. Soc.*, 1985, **107**, 3020; (b) S. Fukuzumi, N. Nishizawa and T. Tanaka, *J. Chem. Soc., Perkin Trans. 2*, 1985, 371.

14. (a) S. Itoh, M. Taniguchi, N. Takada, S. Nagatomo, T. Kitagawa and S. Fukuzumi, *J. Am. Chem. Soc.*, 2000, **122**, 12087; (b) S. Fukuzumi, K. Yasui, T. Suenobu, K. Ohkubo, M. Fujitsuka and O. Ito, *J. Phys. Chem. A*, 2001, **105**, 10501.

15. (a) S. Fukuzumi, T. Okamoto and J. Otera, *J. Am. Chem. Soc.*, 1994, **116**, 5503; (b) S. Fukuzumi, N. Satoh, T. Okamoto, K. Yasui, T. Suenobu, Y. Seko, M. Fujitsuka and O. Ito, *J. Am. Chem. Soc.*, 2001, **123**, 7756.

16. (a) S. Fukuzumi, H. Mori, H. Imahori, T. Suenobu, Y. Araki, O. Ito and K. M. Kadish, *J. Am. Chem. Soc.*, 2001, **123**, 12458; (b) S. Fukuzumi, Y. Fujii and T. Suenobu, *J. Am. Chem. Soc.*, 2001, **123**, 10191.

17. (a) S. Fukuzumi, K. Okamoto and H. Imahori, *Angew. Chem., Int. Ed.*, 2002, **41**, 620; (b) S. Fukuzumi, K. Okamoto, Y. Yoshida, H. Imahori, Y. Araki and O. Ito, *J. Am. Chem. Soc.*, 2003, **125**, 1007; (c) K. Okamoto, H. Imahori and S. Fukuzumi, *J. Am. Chem. Soc.*, 2003, **125**, 7014.

18. (a) S. Fukuzumi, Y. Morimoto, H. Kotani, P. Naumov, Y.-M. Lee and W. Nam, *Nat. Chem.*, 2010, **2**, 756; (b) Y. Morimoto, H. Kotani, J. Park, Y.-M. Lee, W. Nam and S. Fukuzumi, *J. Am. Chem. Soc.*, 2011, **133**, 403; (c) J. Park, Y. Morimoto, Y.-M. Lee, W. Nam and S. Fukuzumi, *J. Am. Chem. Soc.*, 2011, **133**, 5236.

19. S. Fukuzumi, *Prog. Inorg. Chem.*, 2009, **56**, 49.

20. S. Fukuzumi and K. Ohkubo, *Coord. Chem. Rev.*, 2010, **254**, 372.

21. R. I. Cukier and D. G. Nocera, *Annu. Rev. Phys. Chem.*, 1998, **49**, 337.

22. (a) C. J. Chang, M. C. Y. Chang, N. H. Damrauer and D. G. Nocera, *Biochim. Biophys. Acta*, 2004, **1655**, 13; (b) S. Y. Reece and D. G. Nocera, *Annu. Rev. Biochem.*, 2009, **78**, 673.

23. M. H. V. Huynh and T. J. Meyer, *Chem. Rev.*, 2007, **107**, 5004.

24. J. M. Mayer and I. J. Rhile, *Biochim. Biophys. Acta*, 2004, **1655**, 51.

25. J. Rosenthal and D. G. Nocera, *Acc. Chem. Res.*, 2007, **40**, 543.

26. M. M. Pereira, M. Santana and M. Teixeira, *Biochim. Biophys. Acta*, 2001, **1505**, 185.

27. G. W. Brudvig and M. Wikström, in *Photosystem II: The Light-Driven Water:Plastoquinone Oxidoreductase*, ed. T. Wrdrzynski and K. Satoh, Springer, The Netherlands, 2005, pp. 697–713.

28. D. Zaslavsky and R. B. Gennis, *Biochim. Biophys. Acta*, 2000, **1458**, 164.

29. G. T. Babcock, *Proc. Natl. Acad. Sci. U. S. A.*, 1999, **96**, 12971.

30. (a) M. Wikström, K. Krab and M. Saraste, in *Cytochrome Oxidase: A Synthesis*, Academic Press, New York, 1981; (b) S. Ferguson-Miller and G. T. Babcock, *Chem. Rev.*, 1996, **96**, 2889.

31. S. Yoshikawa, K. Shinzawa-Itoh, R. Nakashima, R. Yaono, E. Yamashita, N. Inoue, M. Yao, M. J. Fei, C. P. Libeu, T. Mizushima, H. Yamaguchi, T. Tomizaki and T. Tsukihara, *Science*, 1998, **280**, 1723.

32. (a) T. Chishiro, Y. Shimazaki, F. Tani, Y. Tachi, Y. Naruta, S. Karasawa, S. Hayami and Y. Maeda, *Angew. Chem., Int. Ed.*, 2003, **42**, 2788; (b) J. G. Liu, Y. Naruta and F. Tani, *Angew. Chem., Int. Ed.*, 2005, **44**, 1836.

33. (a) E. E. Chfán, S. C. Puiu and K. D. Karlin, *Acc. Chem. Res.*, 2007, **40**, 563; (b) E. Kim, E. E. Chufán, K. Kamaraj and K. D. Karlin, *Chem. Rev.*, 2004, **104**, 1077.

34. (a) J. P. Collman, R. Boulatov and C. J. Sunderland, in *The Porphyrin Handbook*, ed. K. M. Kadish, K. M. Smith and R. Guilard, Elsevier Science, USA, 2003, vol. 11, pp. 1; (b) J. P. Collman, R. Boulatov, C. J. Sunderland and L. Fu, *Chem. Rev.*, 2004, **104**, 561.

35. F. C. Anson, C. N. Shi and B. Steiger, *Acc. Chem. Res.*, 1997, **30**, 437.

36. S. B. Adler, *Chem. Rev.*, 2004, **104**, 4791.

37. R. Borup, J. Meyers, B. Pivovar, Y. S. Kim, R. Mukundan, N. Garland, D. Myers, M. Wilson, F. Garzon, D. Wood, P. Zelenay, K. More, K. Stroh, T. Zawodzinski, J. Boncella, J. E. McGrath, M. Inaba, K. Miyatake, M. Hori, K. Ota, Z. Ogumi, S. Miyata, A. Nishikata, Z. Siroma, Y. Uchimoto, K. Yasuda, K.-i. Kimijima and N. Iwashita, *Chem. Rev.*, 2007, **107**, 3904.

38. D. T. Sawyer, T. S. Calderwood, K. Yamaguchi and C. T. Angelis, *Inorg. Chem.*, 1983, **22**, 2577.

39. S. Fukuzumi, M. Chiba, M. Ishikawa, K. Ishikawa and T. Tanaka, *J. Chem. Soc., Perkin Trans.*, 1989, **2**, 1417.

40. R. A. Marcus, *Annu. Rev. Phys. Chem.*, 1964, **15**, 155.

41. E. S. Yang, M.-S. Chan and A. C. Wahl, *J. Phys. Chem.*, 1980, **84**, 3094.

42. M. S. McDowell, J. H. Espenson and A. Bakac, *Inorg. Chem.*, 1984, **23**, 2232.

43. S. Fukuzumi, S. Mochizuki and T. Tanaka, *Inorg. Chem.*, 1989, **28**, 2459.

44. (a) S. Fukuzumi, M. Ishikawa and T. Tanaka, *J. Chem. Soc., Chem. Commun.*, 1985, 1069; (b) S. Fukuzumi, M. Ishikawa and T. Tanaka, *Tetrahedron*, 1986, **42**, 1021.

45. K. Matsuura, Y. Yoshii, A. Suzuki and M. Itoh, *Kogyo Kagaku Zasshi*, 1970, **73**, 656.

46. (a) H. Knozinger, *Adv. Catal.*, 1976, **25**, 184; (b) H. A. Benesi and B. H. C. Winquist, *Adv. Catal.*, 1978, **27**, 97.

47. S. Fukuzumi and K. Ohkubo, *J. Am. Chem. Soc.*, 2002, **124**, 10270.

48. S. Fukuzumi and K. Ohkubo, *Chem.–Eur. J.*, 2000, **6**, 4532.

49. R. N. Bagchi, A. M. Bond, F. Scholz and R. Stösser, *J. Am. Chem. Soc.*, 1989, **111**, 8270.
50. W. Känzig and M. H. Cohen, *Phys. Rev. Lett.*, 1959, **3**, 509.
51. H. R. Zeller and W. Känzig, *Helv. Phys. Acta*, 1967, **40**, 845.
52. P. H. Kasai, *J. Chem. Phys.*, 1965, **43**, 3322.
53. J. H. Lunsford, *Catal. Rev.*, 1974, **8**, 135.
54. M. Che, *Chem. Rev.*, 1997, **97**, 305.
55. K. Ohkubo, S. C. Menon, A. Orita, J. Otera and S. Fukuzumi, *J. Org. Chem.*, 2003, **68**, 4720.
56. T. Kawashima, K. Ohkubo and S. Fukuzumi, *Phys. Chem. Chem. Phys.*, 2011, **13**, 3344.
57. S. Fukuzumi, T. Suenobu, M. Patz, T. Hirasaka, S. Itoh, M. Fujitsuka and O. Ito, *J. Am. Chem. Soc.*, 1998, **120**, 8060.
58. S. Fukuzumi, M. Patz, T. Suenobu, Y. Kuwahara and S. Itoh, *J. Am. Chem. Soc.*, 1999, **121**, 1605.
59. S. Fukuzumi, S. Koumitsu, K. Hironaka and T. Tanaka, *J. Am. Chem. Soc.*, 1987, **109**, 305.
60. S. Fukuzumi, K. Okamoto, C. P. Gros and R. Guilard, *J. Am. Chem. Soc.*, 2004, **126**, 10441.
61. K. M. Kadish, L. Frémond, Z. Ou, J. Shao, C. Shi, F. C. Anson, F. Burdet, C. P. Gros, J.-M. Barbe and R. Guilard, *J. Am. Chem. Soc.*, 2005, **127**, 5625.
62. K. M. Kadish, J. Shen, L. Frémond, P. Chen, M. El Ojaimi, M. Chkounda, C. P. Gros, J.-M. Barbe, K. Ohkubo, S. Fukuzumi and R. Guilard, *Inorg. Chem.*, 2008, **47**, 6726.
63. S. Fukuzumi, K. Miyamoto, T. Suenobu, E. Van Caemelbecke and K. M. Kadish, *J. Am. Chem. Soc.*, 1998, **120**, 2880.
64. K. M. Kadish, L. Frémond, J. Shen, P. Chen, K. Ohkubo, S. Fukuzumi, M. El Ojaimi, C. P. Gros, J.-M. Barbe and R. Guilard, *Inorg. Chem.*, 2009, **48**, 2571.
65. S. Fukuzumi, K. Okamoto, Y. Tokuda, C. P. Gros and R. Guilard, *J. Am. Chem. Soc.*, 2004, **126**, 17059.
66. S. Fukuzumi, S. Koumitsu, K. Hironaka and T. Tanaka, *J. Am. Chem. Soc.*, 1987, **109**, 305.
67. S. Fukuzumi, S. Mochizuki and T. Tanaka, *Inorg. Chem.*, 1990, **29**, 653.
68. S. Fukuzumi, Y. Tokuda, T. Kitano, T. Okamoto and J. Otera, *J. Am. Chem. Soc.*, 1993, **115**, 8960.
69. S. Fukuzumi, K. Ohkubo, Y. Tokuda and T. Suenobu, *J. Am. Chem. Soc.*, 2000, **122**, 4286.
70. S. Fukuzumi, K. Shimoosako, T. Suenobu and Y. Watanabe, *J. Am. Chem. Soc.*, 2003, **125**, 9074.
71. (a) C. F. Blanford, R. S. Heath and F. A. Armstrong, *Chem. Commun.*, 2007, 1710; (b) N. Mano, V. Soukharev and A. Heller, *J. Phys. Chem. B*, 2006, **110**, 11180.
72. (a) J. A. Cracknell, K. A. Vincent and F. A. Armstrong, *Chem. Rev.*, 2008, **108**, 2439; (b) I. Willner, Y.-M. Yan, B. Willner and R. Tel-Vered, *Fuel Cells*, 2009, **9**, 7.

73. (a) J. Zhang and F. C. Anson, *J. Electroanal. Chem.*, 1993, **348**, 81; (b) H. Watanabe, H. Yamazaki, X. Wang and S. Uchiyama, *Electrochim. Acta*, 2009, **54**, 1362.
74. M. S. Thorum, J. Yadav and A. A. Grewirth, *Angew. Chem., Int. Ed.*, 2009, **48**, 165.
75. Y. C. Weng, F.-R. F. Fan and A. J. Bard, *J. Am. Chem. Soc.*, 2005, **127**, 17576.
76. K. D. Karlin, S. Kaderli and A. D. Zuberbühler, *Acc. Chem. Res.*, 1997, **30**, 139.
77. S. Fukuzumi, H. Kotani, H. R. Lucas, K. Doi, T. Suenobu, R. L. Peterson and K. D. Karlin, *J. Am. Chem. Soc.*, 2010, **132**, 6874.
78. (a) C. X. Zhang, S. Kaderli, M. Costas, E.-i. Kim, Y.-M. Neuhold, K. D. Karlin and A. D. Zuberbuhler, *Inorg. Chem.*, 2003, **42**, 1807; (b) H. C. Fry, D. V. Scaltrito, K. D. Karlin and G. J. Meyer, *J. Am. Chem. Soc.*, 2003, **125**, 11866.
79. A. A. Gewirth and M. S. Thorum, *Inorg. Chem.*, 2010, **49**, 3557.
80. M. A. Thorseth, C. S. Letko, T. B. Rauchfuss and Andrew A. Gewirth, *Inorg. Chem.*, 2011, **50**, 6158.
81. H. Nagao, N. Komeda, M. Mukaida, M. Suzuki and K. Tanaka, *Inorg. Chem.*, 1996, **35**, 6809.
82. A. J. Bard and L. R. Faulkner, *Electrochemical Methods: Fundamentals and Applications*. 2nd edn, John Wiley & Sons, Inc, New York, 2001.
83. V. G. Levich, *Physicochemical Hydrodynamics*. Prentice-Hall, Inc., Englewood Cliffs, NJ, 1962.
84. J. Koutecky and V. G. Levich, *Zh. Fiz. Khim.*, 1958, **32**, 1565.
85. S. Fukuzumi, T. Kobayashi and T. Suenobu, *J. Am. Chem. Soc.*, 2010, **132**, 11866.
86. S. Fukuzumi, T. Kobayashi and T. Suenobu, *J. Am. Chem. Soc.*, 2010, **132**, 1496.
87. (a) L. Vaska, *Science*, 1963, **140**, 809; (b) L. Vaska, *Acc. Chem. Res.*, 1976, **9**, 175.
88. (a) G. Suardi, B. P. Cleary, S. B. Duckett, C. Sleigh, M. Rau, E. W. Reed, J. A. B. Lohman and R. Eisenberg, *J. Am. Chem. Soc.*, 1997, **119**, 7716; (b) D. B. Williams, W. Kaminsky, J. M. Mayer and K. I. Goldberg, *Chem. Commun.*, 2008, 4195.
89. R. S. Hay-Motherwell, G. Wilkinson, B. Hussain-Bates and M. B. Hursthouse, *Polyhedron*, 1993, **12**, 2009.
90. B. G. Jacobi, D. S. Laitar, L. Pu, M. F. Wargocki, A. G. DiPasquale, K. C. Fortner, S. M. Schuck and S. N. Brown, *Inorg. Chem.*, 2002, **41**, 4815.
91. R. L. Shook, W. A. Gunderson, J. Greaves, J. W. Ziller, M. P. Hendrich and A. S. Borovik, *J. Am. Chem. Soc.*, 2008, **130**, 8888.
92. R. L. Shook, S. M. Peterson, J. Greaves, C. Moore, A. L. Rheingold and A. S. Borovik, *J. Am. Chem. Soc.*, 2011, **133**, 5810.
93. T. H. Parsell, R. K. Behan, M. P. Hendrich, M. T. Green and A. S. Borovik, *J. Am. Chem. Soc.*, 2006, **128**, 8728.

Proton-Coupled Electron Transfer in Natural and Artificial Photosynthesis

M. BARROSO,*[1,2] LUIS G. ARNAUT[2] AND SEBASTIAO J. FORMOSINHO[2]

[1] Department of Chemistry, Imperial College London, South Kensington Campus, SW7 2AZ London, UK; [2] Department of Chemistry, University of Coimbra, Rua Larga, 3004-535 Coimbra, Portugal

5.1 Introduction

Storing solar energy in the form of chemical bonds has been one of the Holy Grails of chemists. High energy demands of modern society, depletion of fossil fuel resources and environmental concerns associated to the increase of anthropogenic CO_2 concentration in the atmosphere, have stimulated a renewed interest for photochemical strategies towards solar energy conversion.

The remarkable system used by nature to convert sunlight into more useful energy equivalents is the source of inspiration for artificial photosynthesis – the development of synthetic systems that mimic the natural photosynthetic processes of light-driven water splitting and CO_2 reduction, for the production of solar fuels.[1–5]

The major bottleneck towards the achievement of efficient artificial photosynthesis is the difficulty in matching the solar irradiance flux with rates of kinetically complex multielectron reactions, which, as seen in Nature, requires

RSC Catalysis Series No. 8
Proton-Coupled Electron Transfer: A Carrefour of Chemical Reactivity Traditions
Edited by Sebastião Formosinho and Mónica Barroso
© Royal Society of Chemistry 2012
Published by the Royal Society of Chemistry, www.rsc.org

catalysts capable of accumulating multiple charges and a good coupling of this process with single electron photochemical events.

In recent years, a significant number of potential catalysts and photocatalysts for water splitting have been proposed,[6–10] many of them inspired by nature's own water oxidation catalyst – the manganese cluster in the oxygen evolving complex (OEC) in Photosystem II. In fact, Photosystem II, present in higher plant and bacteria, is the only known biological enzyme capable of using electrons from water to regenerate its chromophores. The detailed mechanism through which water is oxidised and molecular oxygen is formed in these systems, however, is still subject of debate.[11–15]

Proton-coupled electron transfer (PCET) is a ubiquitous process in biology and chemistry and plays an essential role in multi-electron, multi-proton transfer processes of biological relevance, such as photosynthesis and respiration.[5,17,18] The coupling between proton and electron motions stabilizes reaction intermediates, by preventing charge build-up during the accumulation of redox equivalents. The key role of PCET natural photosynthesis is probably the major contributor to this effect.

In fact, Photosystem II, present in higher plants and bacteria, is the only known biological enzyme capable of using electrons from water to regenerate its chromophores. The detailed mechanism through which water is oxidised and molecular oxygen is formed in biological systems is still subject of debate.[11–15] It is, however, clear that the coupling between proton and electron motions is required to stabilize reaction intermediates, by preventing charge build-up during the accumulation of redox equivalents. These proton-coupled electron transfers (PCET), ubiquitous processes in biology and chemistry, play a vital role in most multi-electron, multi-proton transfer processes of biological relevance.[5,17,18] For this reason it is not surprising that three decades after the first publication on PCET,[16] this is still a highly active area of research.

Nature has devised a structure capable of storing solar energy by splitting water and reducing CO_2 to carbohydrates. To date, however, and despite the significant research effort put into it, synthetic chemistry has not yet been able to reproduce this system and solve that which is one of the most important problems of current chemistry. Understanding the mechanisms of water splitting and O–O bond formation in oxygenic photosynthesis, as well as the impact of chemical modifications in the translation of such mechanisms to synthetic systems, is crucial to the design of efficient artificial photosynthetic cells.

In this chapter, we will start by briefly presenting an historical overview of PCET research, acknowledging the different meanings and mechanistic implications enclosed in the definition and establish the basic principles underlying this type of reactions. Experimental evidence and the importance of PCET in the catalysis of water oxidation in natural photosynthesis is explored, followed by a discussion of how those functional principles may be transposed to synthetic systems, both for fundamental studies and for practical applications in the context of solar energy conversion and solar fuel production.

5.2   Proton-Coupled Electron Transfer Reactions

Although the concept of proton-coupled electron transfer is not new,[16] it has evolved significantly throughout the years and is now used with somewhat different meanings by different research groups. In general terms, PCET can be described as a redox process where both electrons and protons are transferred.

The theory of proton-coupled electron transfer has been developed over the years, with the first theoretical studies published in the 1990s by the groups of Hammes-Schiffer[19–21] and Cukier.[22–25]

The first review on this subject was written by Cukier and Nocera,[26] and several others have been published since, exploring theoretical, experimental and biologically relevant aspects.[9,27–40] Specifically in the context of biological processes, numerous studies have been produced. In particular, the work by Babcock and co-workers on the role of PCET in reaction involving Photosystem II and other enzymes, represents an important landmark in the field.[41–43] More recently, the number of studies focusing on PCET in energy conversion processes and, in particular, in artificial photosynthetic systems, has significantly increased.[36,37,44,45]

Mechanistically, PCET reactions can be classified as concerted (EPT,[33] CEP[18] or CPET[35]), if the transfers of electrons and protons occur in the same reaction step, or stepwise, if electron transfer precedes (ET-PT) or follows (PT-ET) proton transfer. Stepwise mechanisms tend to be energetically more demanding that concerted ones, with higher reaction barriers and slower rates. Concerted reactions avoid the formation of charged species and are important especially in low dielectric media like protein environment. It is therefore not surprising that EPT is preferred in many biological processes, including energy conversion.

The criteria for EPT are that electrons and protons transfer from different orbitals in the donor to different orbitals in the acceptor, and these orbitals are electronically coupled, allowing the events to occur simultaneously. Meyer *et al.* propose an extension of this concept, multiple-site EPT (MS-EPT),[46] where an electron–proton donor transfers to multiple acceptors or an acceptor receives electrons and protons from different donors. These appear to be particularly relevant in biological processes, including the oxidation of tyrosine by P_{680}^+ in oxygenic photosynthesis.

A detailed theoretical description of PCET is outside the scope of this article, and only the main kinetic and thermodynamic aspects will be addressed here.

In the current understanding of PCET reactions, both electron and proton are treated quantum-mechanically, and therefore the tunnelling probability must be accounted for both particles. In fact, concerted processes can be described as double tunnelling (proton and electron), with a single transition state.[26,47] For a description of the reaction coordinate, four adiabatic states (reactants, products and intermediates) described by paraboloids, are usually considered. The expression for the semi-classical rate constant in this case incorporates elements derived from electron and proton transfer theories

$$k_{PCET} = \frac{2\pi}{\hbar} V_{PCET}^2 (4\pi\lambda RT)^{-1/2} \exp\left(-\frac{(\Delta G_{PCET}^0 + \lambda)^2}{4\lambda RT}\right) \qquad (5.1)$$

$$V_{PCET} \approx V_{ET} \times V_{PT} \qquad (5.2)$$

where λ is the reorganisation energy of the system, $\Delta G^0{}_{PCET}$ is the reaction Gibbs free energy, and V_{PCET} represents the product of electron and proton-coupling terms. In addition to the reorganisation energy and reaction driving force, PCET reactions will also depend on the characteristics of the proton transfer process, namely the distance between donor and acceptor as well possibility and frequency of intra- or intermolecular hydrogen bonding. Kinetic isotope effects (KIE), both their magnitude and the effect of pH, can be used as a measure of the role of hydrogen bonding and a tool to assess mechanistic details. In particular, systems where inter- or intramolecular bonds are possible tend to undergo concerted proton and electron transfers, with a weak KIE. The stabilisation of intermediates is compensated by higher reorganisation energy in these cases.

5.2.1 Interfacial PCET

PCET can also play an important role in interfacial charge transfer processes. Electrochemical PCET have recently been explored, both theoretically and experimentally.[48–50] Hammes-Schiffer *et al.* have applied their theoretical framework to model systems where the proton transfer occurs within solvated hydrogen-bonded solute complexes while the electron is transferred between that solute complex and an immersed electrode; Costentin *et al.* have investigated the mechanistic details of PCET in the oxidation of phenols, as model systems of central processes in oxygenic photosynthesis.[35,50–52] The same groups have explored the experimental and theoretical aspects of PCET in model systems with proton relay networks.[53,54]

Another important example of interfacial PCET can occur between semiconductor surfaces and adsorbate molecules,[55] and is particularly relevant for some of the current energy conversion strategies, such as dye-sensitised solar cells (DSSC) or photoelectrochemical (PEC) water splitting cells. A simple proof of the involvement of PCET in interfacial redox processes is the dependence of the conduction and valence band potentials of semiconducting metal oxides, such as TiO_2, with pH.[56] The nature of the surface terminal groups (typically O or OH in metal oxides) will have a strong influence in the thermodynamics and kinetics of the system.

Interfacial electron transfer at solid–liquid interfaces, photoinduced and/or in the presence of an applied potential bias, as in the case of water oxidation on semiconducting metal oxide electrodes involves, as will be discussed in the next section, multiple electron and proton transfer steps. The energy cost associated with charge transfer across the interface will translate into overpotentials for driving the (photo)electrochemical reactions. This is particularly significant in the case of water oxidation on Fe_2O_3 photoelectrodes, for example, where slow surface reactions cannot compete with fast electron-hole recombination.[57] Strategies for optimizing the interfacial process require a good understanding of the mechanistic details, at the molecular level. However, unlike for reactions in solution or biological media, the eventual coupling of electron and proton movement at interfaces has not been significantly explored, either theoretically or experimentally, until very recently.

Petek and co-workers have investigated ultrafast interfacial inner sphere PCET dynamics, where the presence of strong potential gradients will subject electrons and protons to opposite forces within a spatial region.[55] This is the case for water oxidation on semiconductor photoelectrodes, and therefore with potential impact in the context of artificial photosynthesis.

Hammes-Schiffer *et al.* have again extended their theoretical treatment of PCET reactions to investigate the dynamics of PCET at molecule-semi-conductor interfaces.[45] The model systems considered in that work were designed to encompass the cases where electrons are photoexcited from the defect band in the semiconductor and transferred to the adsorbate molecule (as in a typical metal oxide photoelectrochemical water splitting cell), and those where electron excitation occurs in the molecule followed by injection to the conduction band of the semiconductor (the case in dye-sensitised cells). A recent application of this model focused mainly on qualitative charge transfer dynamics, avoiding the complexities inherent to semiconductor-molecule systems, but was capable of providing explanations for the kinetic isotope effects in the population dynamics of CH_3OH/TiO_2, observed by Petek *et al.*[55] Although still incipient, these studies can give an important contribution to the fundamental understanding of interfacial charge transfer processes and ultimately to the development and optimization of semiconductor/dye/catalyst interfaces, which will be discussed in Section 5.4.2.

5.3 Thermodynamics of Water Splitting and CO_2 Reduction

Splitting water into molecular oxygen and hydrogen, and reduction of carbon dioxide into carbohydrates are multielectron processes

$$2H_2O \rightarrow O_2 + 4H^+ + 4e^- \tag{5.3}$$

$$2H^+ + 2e^- \rightarrow H_2 \tag{5.4}$$

$$nCO_2 + 2nH^+ + 2ne^- \rightarrow (CH_2O)_n \tag{5.5}$$

and, as such, are intrinsically complex, both kinetically and thermodynamically.

In the case of water oxidation to molecular oxygen (eqn (5.3)), the minimum energy path corresponds to a series of four one-electron/one-proton steps, each of which requires an average energy of 0.81 V *versus* NHE at pH 7 (Table 5.1). This ideal situation is depicted by the dotted line in the Frost diagram in Figure 5.1a. In reality, the reaction pathway will depend on the specific conditions of the system, and it can involve significant variations in the energy requirement for each individual step. This is exemplified by the dashed line in Figure 5.1a where the first step, the one-electron/one-proton oxidation of H_2O to $^{\cdot}OH$, requires 2.33 eV, followed by the formation of H_2O_2 with a formal redox potential of only 0.38 V *versus* NHE. The importance of coupling electron to proton transfers in this context can be understood by comparing the thermodynamic barriers for the one-, two- and four-electron oxidation of water, that correspond to the initial slopes of the dotted, full and dashed line, respectively, in Figure 5.1a.

Similar considerations can be made for the evolution of H_2 from proton reduction (Figure 5.1b, Table 5.1) or reduction of CO_2 to methane or methanol (Figure 5.1c, Table 5.1). In all cases illustrated, the minimum energy pathways are those that involve the transfer of multiple electrons. However, the thermodynamic advantage associated to the minimum energy paths is countered by mechanistic difficulties, and in general, such reactions can only be effected efficiently with the assistance of suitable catalysts. Natural photosynthetic systems are remarkable in that they incorporate efficient catalytic units that are capable of promoting the mechanisms of water oxidation and reduction that are closer to the minimum energy paths. This is the case of the oxygen-evolving centre (OEC) in Photosystem II (PSII) of higher plants, and hydrogenases coupled to Photosystem I (PSI) in some algae and cyanobacteria. The specific characteristics of these systems, and in particular the importance of coupling

Table 5.1 Selected single- and multielectron formal reduction potentials for water oxidation and proton and CO_2 reduction reactions, in water at pH 7. (P. M. Wood, *Biochem. J.*, 1988, **253**, 287; E. Fujita, *Coord. Chem. Rev.*, 1999, **185–186**, 373.)

Reaction	E°/V vs. NHE (pH 7)
$OH^{\cdot} + H^{+} + e^{-} \rightarrow H_2O$	2.33
$H_2O_2 + 2H^{+} + 2e^{-} \rightarrow 2H_2O$	1.35
$O_2 + 4H^{+} + 4e^{-} \rightarrow 2H_2O$	0.82
$H^{+} + e^{-} \rightarrow H^{\cdot}$	-2.20
$2H^{+} + 2e^{-} \rightarrow H_2$	-0.41
$CO_2 + e^{-} \rightarrow CO_2^{\cdot -}$	-1.90
$CO_2 + H^{+} + 2e^{-} \rightarrow HCO_2H$	-0.61
$CO_2 + 2H^{+} + 2e^{-} \rightarrow CO + H_2O$	-0.53
$CO_2 + 4H^{+} + 4e^{-} \rightarrow HCHO + H_2O$	-0.48
$CO_2 + 6H^{+} + 6e^{-} \rightarrow CH_3OH + H_2O$	-0.38
$CO_2 + 8H^{+} + 8e^{-} \rightarrow CH_4 + 2H_2O$	-0.24

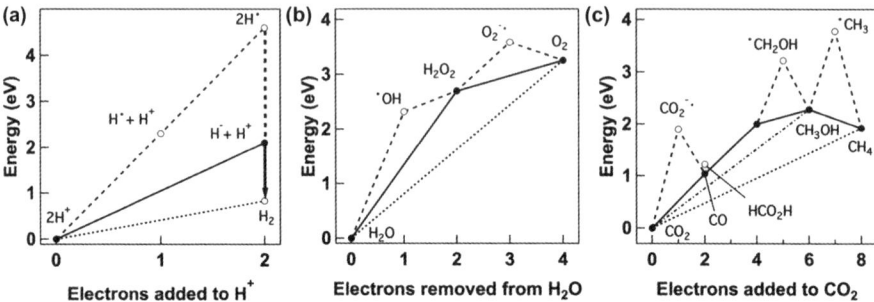

Figure 5.1 Modified Frost diagrams for (a) hydrogen, (b) oxygen and (c) carbon, illustrating different pathways for oxygen production from water oxidation, hydrogen evolution and CO_2 reduction, at pH 7 *versus* a Normal Hydrogen Electrode (NHE) reference.

the motions of protons and electrons in the catalytic process, will be discussed in the following sections.

5.4 Natural Photosynthesis

Photosynthesis is the mechanism that allows plants, algae and bacteria to store solar energy in the form of chemical bonds, through a complex sequence of carefully coupled events, including light harvesting, energy and charge transfer, spatial separation of charges and synthesis of the chemical compounds that will ultimately store part of the light energy absorbed,

$$6H_2O + 6CO_2 \xrightarrow{hv} C_6H_{12}O_6 + 6O_2 \quad (E^\circ = 1.24 \text{ V}) \tag{5.6}$$

Although a great variety of photosynthetic systems can be found in nature, reflecting a multitude of different environments, the underlying principles of solar energy conversion are generally common to the majority of the systems.

Light absorption is mainly attained by arrays of chromophoric molecules that function as antennas, trapping photons over a wide wavelength range and funnelling this photonic energy to the reaction centre.

In oxygenic photosynthesis, nicotinamide adenine dinucleotide phosphate ($NADP^+$) is reduced to NADPH, with electrons supplied from water oxidation at the Mn_4Ca cluster in the oxygen-evolving complex (OEC). The energy input for the overall redox process is obtained through light excitation of the central chlorophylls in two reaction centres, Photosystems I (PSI) and II (PSII), present in the thylakoid membranes of chloroplasts. Additional redox cofactors, such as cytochrome b_6f, are then responsible for the transport of electrons from PSII to PSI. This process is coupled to proton movement, creating a proton gradient across the thylakoid membrane that drives the conversion of

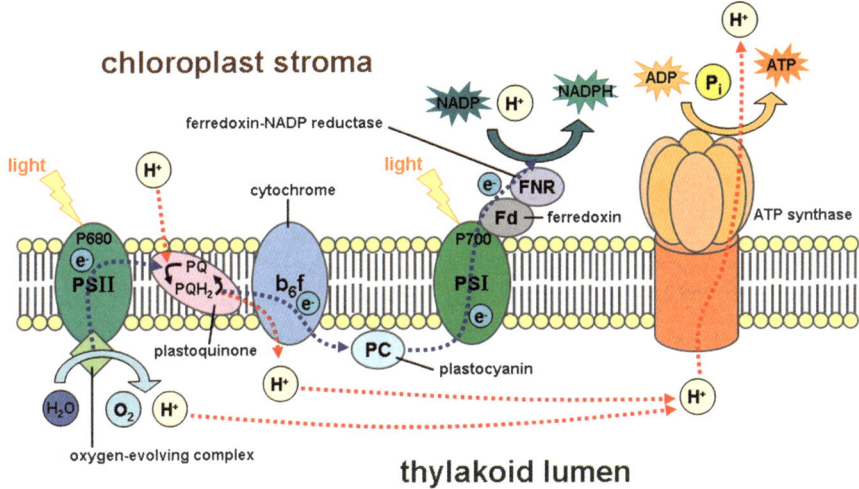

Figure 5.2 Illustration of the arrangement of the cofactors involved in oxygenic photosynthesis in a chloroplast. Main electron and proton transfer processes are indicated by dashed blue and red arrows, respectively. (Source: Wikipedia)

ADP to ATP, as illustrated in Figure 5.2. The reduction of CO_2 and production of sugars occurs *via* light-independent reactions, in the Calvin Cycle, where NADPH and ATP are finally consumed.

In nature, Photosystem II is the only enzyme capable or regenerating its oxidized chromophores using water as an electron donor. For this reason, a detailed understanding of the energetic and mechanistic aspects of light induced reactions in PSII, particularly water oxidation, is extremely appropriate and valuable for the design of artificial systems for solar fuel production, and will be discussed in the following sections.

5.4.1 Structure and Mechanism of Photosystem II

PSII has been extensively investigated and a great deal of information is now available regarding its reactivity,[12,15,43] mechanism[14,58–61] and structure.[62–64] Figure 5.3 depicts the structure and arrangement of the main cofactors of PSII in cyanobacteria *Thermosynechococcus elongatus*,[63] where the sequence of electron transfer processes relevant for water oxidation is indicated with arrows. The overall redox process can be described as

$$2H_2O + 2Q_B + 4h\nu \rightarrow O_2 + 2Q_BH_2 \qquad (5.7)$$

where the oxidation of water to O_2 is accompanied by the reduction of plastoquinone Q_B to plastoquinol, Q_BH_2, and subsequent release of four proton and four electrons in the thylakoid membrane of the chloroplast.

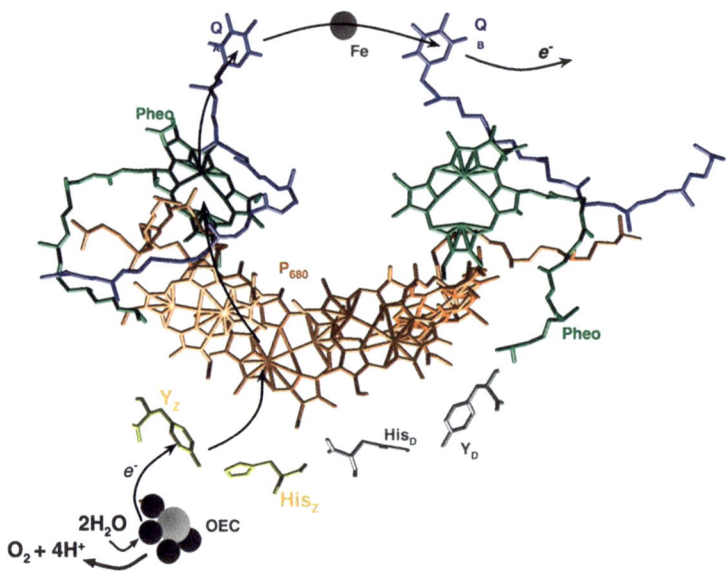

Figure 5.3 Scheme of the structure of PSII with main cofactors and electron transfer processes involved in water oxidation. From the X-ray structure of cyanobacteria *Thermosynechococcus vulcanus*, with a resolution of 2.9Å (X-ray structure PDB:3BZ1).[63]

Light excitation of chlorophylls in P_{680}, the primary electron donor in PSII, is followed by long-range ultrafast electron transfer (~ 200 ps) from P_{680}^* to the quinone acceptor Q_A in the acceptor side, *via* a pheophitin (Pheo) bridge, forming a charge separated state, $P_{680}^+PheoQ_A^-$ which is stable for 100–200 μs. This process is followed by charge transfer from Q_A^- to Q_B, and, upon a second turnover of P_{680}, a second electron is transferred to Q_B^- with concomitant protonation, forming the hydroquinone H_2Q_B. The hydroquinone will then be released in the thylakoid membrane to deliver electrons to PSI, and will be replaced by a new Q_B unit.

The estimated redox potential of P_{680}^+ is *ca.* 1.25 V *vs.* NHE[11,65] and therefore sufficient to oxidize water ($E^{o'} = 0.9$ V *vs.* NHE at pH $= 5.5$). This process is mediated by the tyrosine residue Y_Z, that oxidizes P_{680}^+ in the donor side of PSII within nanoseconds of light induced charge separation, preventing recombination of the charge separated $P_{680}^+–Q_A^-$ and increasing spatial separation of oxidizing and reducing equivalents (Figure 5.4).[15,66]

Tyrosine Y_Z is also an important mediator in the catalysis of water oxidation by the Mn_4Ca cluster in the OEC. This proceeds *via* a series of proton-coupled electron transfer events, each of which reduces one radical $Y_Z^\bullet$ and increases the oxidation state of the cluster. The different oxidation states of the OEC are known as S-states. Once state S_4 is attained, corresponding to the accumulation of four electron holes, two water molecules are oxidised, and a molecule of oxygen is formed. Kok proposed a S-state cycle describing the successive

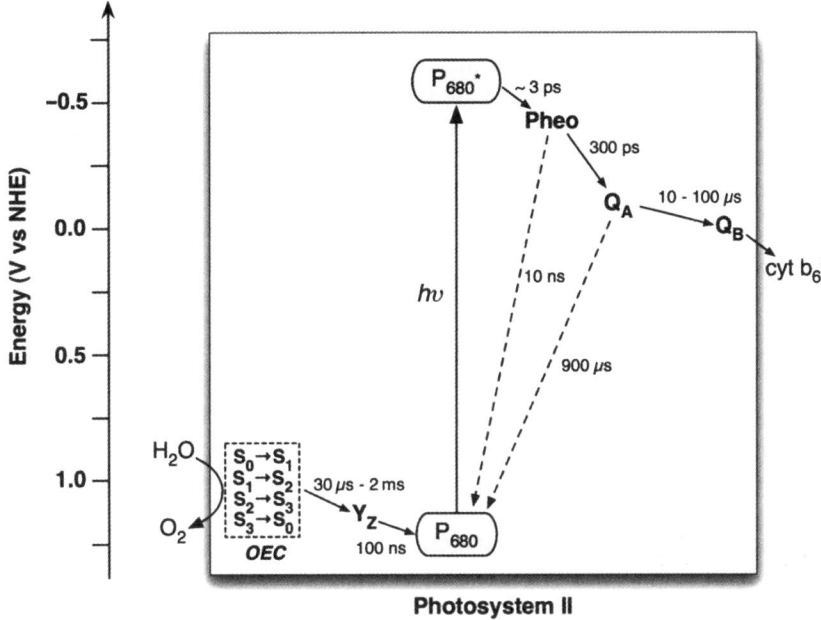

Figure 5.4 Energy level diagram and timescales of charge transfer events following light excitation in photosystem II.

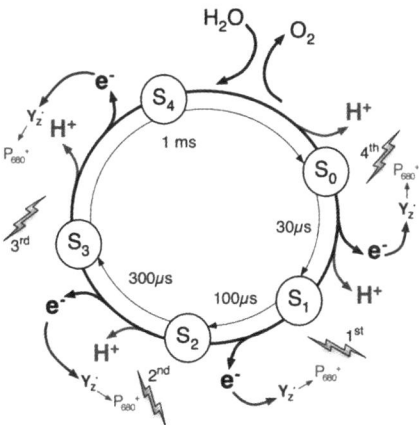

Figure 5.5 Modified S-state (Kok) cycle to include proton and electron transfer steps. Adapted from ref. 77.

oxidation of the manganese cluster.[67] As more structural and mechanistic details of the OEC are unveiled, modifications to the original Kok cycle have been proposed[11,68] (Figure 5.5).

Nature has designed a remarkably efficient system where successive electron transfer reactions following light excitation quickly yield stable charge separation with minimal energy losses, while coupling multi-redox catalysis with single photon processes, but also equipped with self-protection and regeneration mechanism. The role of PCET in the proton management and redox leveling of the intermediates formed is essential for the efficiency displayed by PSII, which is capable of producing up to 50 molecules of O_2 per second.[14,58,60,69] In the next section, the function of tyrosine and the catalysis of water oxidation by the OEC will be discussed.

5.4.2   PCET in Photosystem II

The identification of tyrosine-161 of the D1 subunit in PSII (Y_Z), as the redox mediator between the oxidation of the OEC and reduction of $P_{680}{}^+$, was first proposed by Babcock *et al.*[41–43,70] and Britt *et al.*[71] Krishtalik had previously suggested that such a basic species, capable of binding the protons released from water and with a pK_a dependent on the oxidation state of the manganese cluster, would be required to stabilise the reaction intermediates,[72,73] while Witt *et al.*[74] identified that species as a tyrosine from the results of EPR studies.

Tyrosine Y_Z is an amino acid with a phenolic group that becomes very acidic when Y_Z is oxidized and leads to the deprotonation of the radical to a histidine residue (His 190) to which it is hydrogen bonded (Scheme 5.1). The exact transfer mechanism is still subject of debate and has been explored by several groups either for natural or synthetic model systems.[31,32,50–52,75,76]

The most recent structure of the PSII with a resolution of 1.9Å was recently presented for the cyanobacteria *Thermosynechococcus vulcanus*.[64] New features have emerged from this study, particularly in regard to the structure and organization of the OEC cluster, the presence of substrate water molecules and the pattern of hydrogen bonding not observed in previous studies. The manganese cluster, Mn_4CaO_5 is described as a cubane, as in previous studies[62] but slightly distorted, with three of the manganese, four oxygen atoms and the Ca atoms on the corners. The fourth Mn is outside the cubane, attached to an oxygen in the cubane corner and to the fifth oxygen by μ-oxo bridges. Additionally, due to the

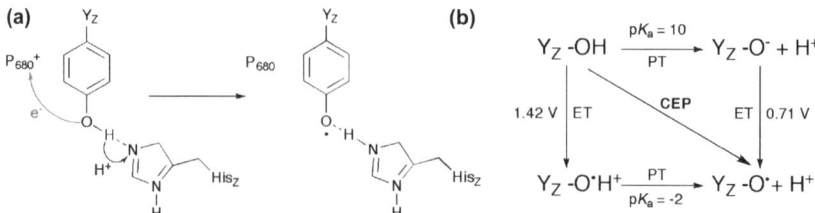

Scheme 5.1   (a) Concerted bidirectional proton and electron transfer in of tyrosine in PSII. (b) Square diagram illustrating the thermodynamic requirements associated to stepwise and concerted mechanisms.

presence of water molecules in close proximity, a saturating ligand environment is observed for Mn_4CaO_5, with each Mn attached to six ligands and the Ca attached to seven ligands. Shen and co-workers suggest that this arrangement may have an important role in the mechanism of water oxidation and O=O bond formation. Finally, an extensive hydrogen-bonding network was found between Y_Z and the manganese cluster. In particular Y_Z appears to establish short strong hydrogen bonds with a water molecule connected to the cluster, and with His190. This hydrogen-bond network is believed to function as a proton exit channel, supporting oxidation of Y_Z by PCET.

Although much progress has been made in respect to understanding the mechanisms of the OEC, the detailed mechanism of water oxidation and O_2 bond formation is not fully understood. It is believed that O=O bond formation proceeds *via* high valent Mn intermediate species, such as Mn^V=O or Mn^{IV}–O·.[13,59] As discussed before, the sequential oxidation of Mn atoms in the OEC occurs on the microsecond to millisecond timescale (Figure 5.4).[77] This is possible due to the levelling of the redox potentials associated with each of the steps in the Kok cycle.[65] Proton-coupled electron transfer is believed to be responsible for the redox levelling in the OEC.[13,78] As the valence state of the manganese cluster increases, water molecules are deprotonated, stabilising the entire molecular ensemble in this way.

5.5 Artificial Photosynthesis

A detailed understanding of the mechanisms behind light driven water splitting in nature is essential for the guided design of new assemblies that can mimic and ultimately exceed the performance of biological organisms. Like their natural analogues, efficient artificial photosynthetic systems must be capable of (i) capturing light in the visible region of the solar spectrum, (ii) producing charge separated states with minimum energy losses, and (iii) achieving multi-electron, multi-proton catalysis.[2,79,80] The first two criteria are reasonably easy to verify with the current level of understanding of photochemical systems and characterization techniques available. Light harvesting units have been proposed that emulate the role of antennas in PSII,[81,82] and several donor-acceptor assemblies have been shown to achieve efficient energy transfer, as well as spatial and temporal charge separation, comparable to those in natural photosynthesis.[83,84] Matching one electron photo-processes with multi-redox catalysis, however, is a considerable challenge and man-made systems are yet to reach the performance and stability of the OEC.

Light induced water splitting using semiconductor-based photoelectrodes,[85,86] nanoparticulate metal oxides,[86–89] (oxy)nitrides and (oxy)sulfides,[90,91] and molecular catalysts that mimic aspects of the active sites in natural enzymes[10,92–96] have been extensively investigated and a wealth of new materials and device architectures has been presented in recent years. In the following sections we will look in detail at examples of integrated systems inspired in Photosystem II and how PCET can play a key role in the corresponding reaction mechanisms.

5.5.1 Model Systems for Photosystem II

The development of model systems can make a significant contribution towards the progress of artificial photosynthesis by simultaneously providing information relevant for an increased understanding of the working principles of natural systems, and allowing a systematic approach to the synthesis of bio-inspired systems that can be incorporated in working devices.

Ideally, a molecular model of PSII should include (i) a photoactive unit that efficiently absorbs visible light energy (S), (ii) an electron donor (D), (iii) an electron acceptor (A), (iv) an efficient electron relay between donor and acceptor units to induce directionality in electron transfer and efficient charge separation, and finally (v) catalytic units capable of accumulating multiple oxidising equivalents and promoting the oxidation and reduction of water (WOC and WRC). This model system is illustrated in Figure 5.6.

Numerous architectures have been proposed for the A–S–D component of the photosystem model, that exhibit efficient electron transfer cascade processes resulting in long-lived charge-separated states. The simplest of such systems are triads, but tetrads and pentads have also been reported.[83,97–100] Due to its robustness and unique photophysical and electrochemical properties, $[Ru(bpy)_3]^{2+}$ is commonly the sensitiser of choice for most of these systems. Excitation of this chromophore with light in the region around 450 nm induces a metal to ligand charge transfer (MLCT) resulting in an excited state that lives long enough to undergo subsequent reductive or oxidative electron transfer reactions. Another important reason for the choice of this sensitiser is the oxidation potential of the cation $[Ru(bpy)_3]^{3+}$, around 1.2 eV, which is very close to that of P_{680}^+, as discussed earlier. Furthermore, modifications in the structures of ligands can be used to modulate the photophysical and electrochemical properties of the ruthenium complex, making it very attractive for utilization in systematic studies. As we will see later, the remarkable qualities of ruthenium polypyridyl complexes mean that its application in the context of solar energy conversion is not limited to sensitisation, but can also encompass a catalytic function.

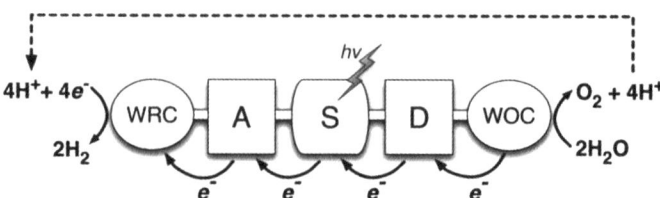

Figure 5.6 Schematic representation of a model artificial photosynthetic system.

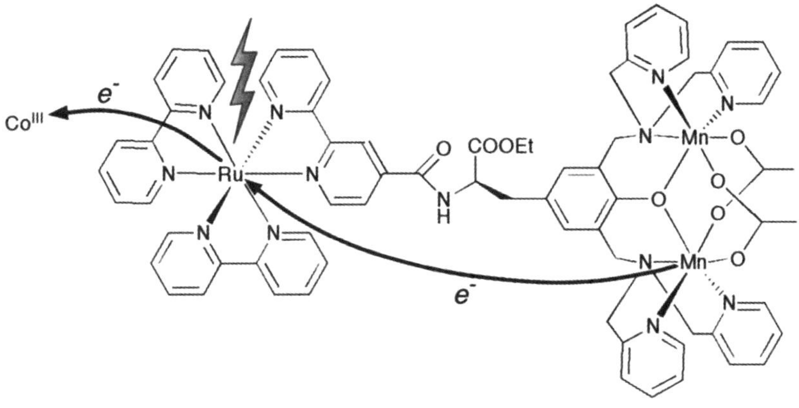

Figure 5.7 Model system including a ruthenium photosensitiser unit connected to a tyrosine and a dimer manganese unit, proposed by Hammarstrom *et al.* to demonstrate light-induced accumulative electron transfer.[18]

On the donor side of PSII, as mentioned previously, tyrosine Y_Z reduces P_{680}^+ and simultaneously transfers a proton to hystidine His190, to which it is hydrogen-bonded. This central process in oxygenic photosynthesis has been extensively investigated and modelled, both for obtaining insights into the working principles of PSII and to develop synthetic systems that feature similar functionalities, in particular PCET as a means of proton management and redox levelling.

The first studies of a synthetic system with a tyrosine bound to a $[Ru(bpy)_3]^{2+}$ photosensitiser were performed in water, in the presence of methyl viologen (MV^{2+}) as an electron acceptor.[101] Irradiation of the complex with a 460 nm laser pulse resulted in the bleach of the MLCT band due to the oxidation of the Ru(ii) to Ru(iii), with concomitant reduction of the electron acceptor to $MV^{\bullet+}$. The shorter recovery time of the chromophore compared to the lifetime of $MV^{\bullet+}$ suggested an intramolecular PCET from the tyrosine to Ru(ii). The observation of the tyrosyl radical by EPR spectroscopy provided the confirmation of the similarity between the mechanisms in the synthetic system and P_{680}^+-Y_Z in PSII. A later study demonstrated that the analogy between the synthetic and natural system could be taken further, when the tyrosyl radical formed was shown to oxidise a Mn(iii) dimer complex,[102] as $Y_Z^{\bullet}$ is capable of oxidizing the manganese cluster in the OEC. These two examples confirm the importance of PCET in the redox processes involving tyrosine, and demonstrate the possibility of incorporating PCET in artificial photosynthetic systems (Figure 5.7).

5.5.2 Water Oxidation Catalysts

The mutielectron oxidation of water requires a suitable catalyst, that is capable of accumulating multiple oxidising equivalents and catalysing the breaking of four O–H bonds and the formation of a O=O bond, from two water molecules.

A number of oxidation catalysts have been investigated in the context of solar water splitting, from crystalline metal oxides, to molecular transition metal complexes.[7,85,87] Common to most of these is the presence of one or multiple transition metals centres, as in the OEC in PSII.

5.5.2.1   Solid State Catalysts

Heterogeneous water oxidation catalysts are chemically rugged and in some cases show self-repair properties, similar to the case with the calcium-manganese cluster in OEC.

The best-known examples of heterogeneous water oxidation electrocatalysts are IrO_2,[87,88] Co_3O_4,[87,89,103] and RuO_2,[87] which have been studied since the early 1970s and show the lowest overpotentials for water oxidation. More recently, new heterogeneous structures have been proposed that show equally promising catalytic performances. The use of cobalt based catalysts, for example, has been explored by several different groups, with encouraging results. One such example is the cobalt-phosphate (CoPi) catalyst reported by Kanan and Nocera.[68,104–106] This is perhaps one of the most discussed heterogeneous catalyst structures in recent years, due to its apparent similarities to the OEC manganese cluster: (i) cubane structure, (ii) self-assembly, self-repair, (iii) operation at neutral pH. Also important is the fact that it is made from cheap and abundant materials and can be operated in non-purified water. Furthermore, EXAFS and XANES studies revealed that the high-valent intermediate Co^{IV}-O is involved in the O–O bond formation step and appears to be formed from Co^{III}-OH through a proton-coupled electron transfer step.[107] The role of phosphate in the catalyst structure has been discussed and it was initially thought to intervene much in the same way as histidine in PSII, accepting the proton released from water, as a proton acceptor. More recent studies, however, showed that replacing phosphate by other conjugated bases would produce similar results, therefore contradicting the initial mechanism.[108] To obtain a better insight into the mechanistic details of the cobalt cubane catalyst, Nocera *et al.* isolated the kinetics of the PCET formation of Co^{IV}-O by studying an analogous cubane complex $[Co_4O_4(CO_2Me)_2(bpy)_4](ClO_4)_2$. The effect of pH in the electrochemical response of the complex and the magnitude of the KIE observed, allowed the establishment of two different PCET mechanisms: bidirectional PCET at an electrode surface and unidirectional concerted proton electron transfer for the self exchange reaction.[109]

Another recent example of an heterogeneous OEC analogue is the $[Mn_4O_4]^{6+}$ cubane, proposed by Dismukes *et al.*[110,111] Also in this case it has been suggested that the similarities with the OEC include the capacity of the catalyst to self assemble and regenerate. This catalyst can oxidise water with sunlight as the sole energy source, when a $Ru(bpy)_3^{2+}$ dye is coupled as light harvester.

5.5.2.2   Homogeneous Catalysts

Homogeneous water oxidation catalysts are easily dispersed and their properties can generally be tuned by ligand modification. Such properties make these systems highly attractive for the purposes of systematic design of new optimized structures and mechanistic investigation.

The calcium-manganese cluster in PSII-OEC has inspired the design of a number of synthetic catalysts. Like in the natural system, the majority of the artificial oxygen evolution centres contain one or multiple transition metal atoms, the most common being Mn, Ru and Ir (Chart 5.1). These transition metals are known for their ability to undergo multiple oxidations, while the higher oxidation states can be stabilized by the introduction of electron donating groups.

Several molecular homogeneous catalysts are currently available, with well-characterised structures and reasonably well understood operating mechanisms.[10]

The first known example of a molecular catalyst, capable of promoting the evolution of oxygen from water is the "blue dimer", $cis,cis[(bpy)_2(H_2O)$-$Ru^{III}ORu^{III}(H_2O)(bpy)_2]^{4+}$, (**4**, Chart 5.1, bpy = 2,2'-bipyridine), reported by

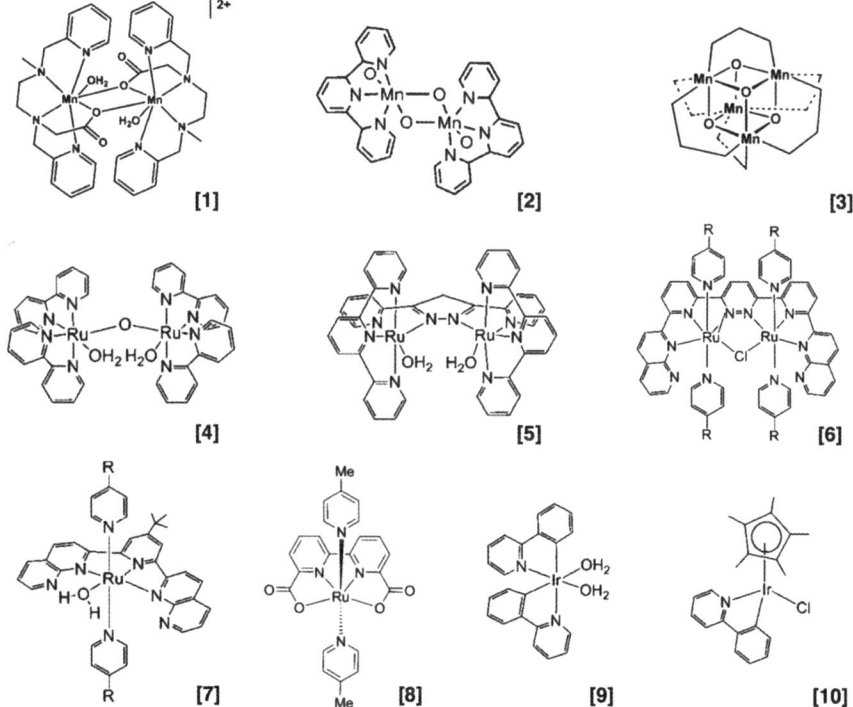

Chart 5.1 Selected homogeneous oxidation catalysts: [1] $[Mn_2(pmpa)_2(\mu\text{-}O)_2]^{n+}$, pmpa = bis(2-pyridylmethyl)amino)propionic acid; [2] [(terpy)(H_2O)-Mn(μ-O)_2 Mn(terpy)(H_2O)]; [3] $Mn_4O_4L_6$ cubane; [4] "blue dimer", *cis,cis*-$[(bpy)_2(H_2O)Ru^{III}ORu^{III}(H_2O)(bpy)_2]^{4+}$; [5] Ru-Hbpp; [6] pyridazine-bridged Ruthenium dimer complexes; [7] $[\mathbf{Ru}^{II}\text{-}OH_2]^{2+}$ with **Ru** = Ru(NPM)(pic)_2, NPM = 4-*t*-butyl-2,6-di-(1',8'-naphthyrid-2'-yl)-pyridine, pic = 4-picoline [8] Ru(II)L(pic)_2, H_2L = 2,20-bipyridine-6,60-dicarboxylic acid and pic = 4-picoline; [9] $[Ir^{III}(L)_2(H_2O)_2]^+$, L = 2-(2-pyridyl)phenylate; [10] $[Cp^*Ir(ppy)Cl]$.

Meyer in 1982.[112,113] In the presence of Ce(IV) as an oxidizing agent, an average of 13 turnovers and a rate of 0.0042 O_2 molecules evolved per second, has been reported. The work represented a breakthrough in the field of artificial photosynthesis and the ruthenium dimer has been extensively investigated since then, in an attempt to obtain insights into the mechanism for water oxidation.[114-122] Some mechanistic details are still subject to debate, largely due to the difficulty in directly probing the intermediate states. It is, however, accepted that the ruthenium dimer undergoes oxidative activation by stepwise proton-electron loss, producing the reactive high-valent intermediate $[(bpy)_2(O)Ru^VORu^V(O)(bpy)_2]^{4+}$. This intermediate is then attacked by a water molecule, forming a peroxido intermediate that is subsequently oxidized, releasing O_2 and regenerating the catalyst.[117] The role of PCET in this mechanism is critical. By avoiding charge build-up, it allows the accumulation of multiple oxidizing equivalents and the stabilization of high-energy intermediates.[123]

The importance of PCET and concerted proton electron transfer (EPT) in the activation of ruthenium oxido complexes was initially acknowledged for $[Ru(bpy)_2(py)OH_2]$, but this complex failed to catalyse water oxidation.

The failure of the mononuclear $[Ru(bpy)_2(py)OH_2]$ to promote oxygen evolution followed by success of the blue dimer and the fact that the OEC itself contains a multi-metal centre, lead researchers to believe that multiple sites were required for catalytic water oxidation. However, the mechanism proposed for the activation of the blue dimer showed that in reality, only one of the metal centres is involved in the oxygen-oxygen bond formation step. This suggestion was later confirmed when Meyer and co-workers showed that single site ruthenium-aqua complexes were capable of oxidizing water in the presence of Ce(IV). One of these catalysts, $[Ru(tpy)(bpm)OH_2]^{2+}$ (tpy = terpyridine, bpm = 2,2'-bipyrimidine), is shown in Scheme 5.2.

Ruthenium is undoubtedly the most used transition metal in synthetic molecular water oxidation catalysts at present, mostly due to the robustness of the metal-ligand bonds formed, which increase the stability of the complexes in higher oxidation states. However, these catalysts present other limitations, in particular the possible involvement of kinetically inert intermediates in the water oxidation process, which necessarily decreases the turnover of these systems.

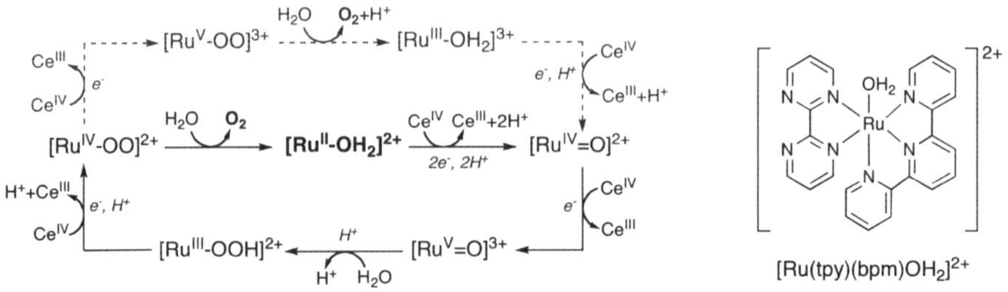

Scheme 5.2 Proposed mechanism for the oxidation of water by the single site catalyst $[Ru(tpy)(bpm)(OH_2)]^{2+}$ (Ru^{II}-OH_2 in the scheme).

In nature, manganese is the metal of choice, which has inspired the development of numerous Mn-based molecular catalysts, in particular complexes including Mn atoms linked by μ-oxo bridges, as in the OEC. Examples of oxomanganese complexes are illustrated in Chart 5.1 ([1]-[3]). A wide range of mono-, di- and tetra-manganese complexes have been described in the literature as capable of oxidising water. However, the exact provenance of oxygen in the evolved O_2 is not always clear, as these complexes are very labile and tend to decompose after a few turnovers. Among the most stable catalysts, the ones based on [Mn_2O_2] cores are the most promising, with catalytic rates in the order of 2×10^{-3} s^{-1} molecules of O_2 evolved per second.[124]

Given the high catalytic activity of IrO_2, several molecular iridium complexes have also been presented in the literature in the past few years. Among them, [$Ir^{III}(L)_2(H_2O)_2$]$^+$, with L = 2-(2-pyridyl)phenylate anion[125] ([9], Chart 5.1) and analogue ligands, have been reported to catalyse water with remarkably high turnover numbers (>2500) in the presence of Ce^{IV} as oxidising agent. Further characterisation, including kinetic and mechanistic studies is still required particularly to understand the nature of the intermediate states involved and the interaction between proton and electron motion in the catalytic process.

The functionalization of semiconductor materials with water oxidation catalysts is another promising solution to achieve charge separation and catalytic function with robust assemblies. Significant enhancement of photocurrent response has been observed with WO_3 and Fe_2O_3 catalyst functionalised with the cobalt based oxygen-evolving catalyst discussed previously.[126–130] Suggestion of the involvement of PCET also in water oxidation in these systems have been advanced, but to date no detailed mechanistic studies have been performed in order to clarify this hypothesis.[131]

A further extension of the previous approach consists of coupling water oxidation catalysts to photochemical units creating complex photoelectrochemical nanostructures that mimic biologic proton-coupled electron transfer in Photosystem II.[132]

Youngbood *et al.*, have shown, for the first time, a photoelectrochemical cell inspired by PSII, where light harvesting, charge separation and water oxidation are achieved in separate units[133] (Scheme 5.3).

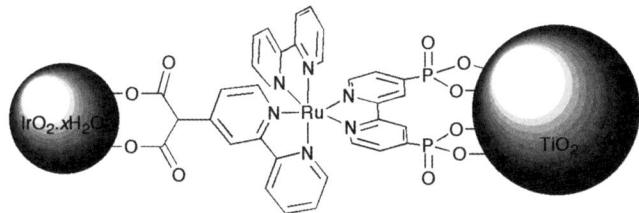

Scheme 5.3   Illustration of the main units (electron acceptor, photosensitiser and water oxidation catalyst) in the dye-sensitised water splitting cell proposed by Youngblood *et al.*[133]

The modular approach[45] discussed above clearly represents an engineering challenge. Regardless of the individual performance of the light harvesting, charge separation and catalytic modules, the kinetics and thermodynamics of the corresponding processes must be matched, to guarantee an overall efficient process.

One possible alternative to the complex modular scheme has been proposed by Meyer and co-workers, and is based upon the integration of excitation, electron, and proton transfer in one single concerted step. Instead of conventional PCET, where light excitation precedes a proton and electron transfer event, the authors propose a mechanism based on *photo-EPT*. This as a concept has been tested for intramolecular charge transfer in hydrogen bonded dyes, but further development is needed before it can be applied to light driven water splitting devices.[134,135]

5.5.3 Proton and CO_2 Reduction

So far we have focused our discussion on the water oxidation half-reaction of photosynthesis, to illustrate the importance of PCET in natural and artificial solar energy conversion systems. A similar discussion is pertinent in the context of the reduction counterparts, where solar fuels like H_2, CH_4 or CH_3OH are effectively produced.

As discussed in section 5.3 (eqn (5.4) and (5.5), Table 5.1), the evolution of molecular hydrogen from water splitting and the reduction of CO_2 to hydrocarbons involve the transfer of multiple electrons. In particular, light driven reduction of CO_2 to methane or methanol with concomitant water oxidation, is kinetically very demanding due to the involvement of four or six proto/electron transfers, respectively. Additionally, the one electron reduction of CO_2 to CO_2^- is thermodynamically unfavourable ($E° = -1.9$ *vs.* NHE at pH 7). The development of bio-inspired reduction catalysts that lower the activation barrier and facilitate such reactions is currently being pursued by several research groups and a number of important achievements have been reported in recent years.[79,96,136]

Light driven reduction of CO_2 to methane or methanol with concomitant water oxidation, as practical advantages over hydrogen evolution, since discussed in section 5.2, is kinetically very demanding, due to the involvement of four or six proto/electron transfers, respectively. Additionally, the one electron reduction of CO_2 to CO_2^- is thermodynamically unfavorable ($E° = -1.9$ *vs.* NHE at pH 7).

Most electrochemical and photoelectrochemical systems produce only the two-electron reduction products of CO and formate, the products of two-electron reductions, evidencing once again the kinetic bottleneck associated wth multielectron and multiproton processes.[137-139] As in the oxidation of water, efforts have been directed towards the development of transition-metal based electrocatalysts with multiple metal centres to facilitate charge accumulation in highly reduced intermediates and allow multiredox processes to occur.[140-142]

These catalysts, however, have failed to promote CO_2 reduction involving more than two electrons.

Currently there are a number of reports in the literature of electro-catalysts[143–150] and photocatalysts[151–153] capable of reducing CO_2 to methanol, but there has been limited effort towards the understanding of mechanistic details,[150,152,154] and further work is needed to assess the role of PCET in these systems and devise structures capable of reducing CO_2 to methane or methanol.

5.6 Concluding Remarks

From the studies on photosystem II and synthetic functional models discussed here, three main aspects have emerged as central for water oxidation: (i) accumulation of multiple oxidising equivalents by transition metal oxidation, (ii) redox potential levelling by avoiding charge build-up in intermediates, and (iii) efficient proton management and proton removal before O–O bond formation. PCET accounts for the last two criteria and plays a crucial role throughout the entire photosynthesis process.

In particular for the case of water oxidation, lessons learned from natural and bio-inspired systems allow us to establish a set of guidelines for the design of efficient catalysts: (i) the catalyst must be able to bind to the substrate, water, in order to facilitate de removal of protons, (ii) it must be capable of under-going multiple oxidation steps (S_0–S_4) within a narrow potential range, (iii) it must be stable in at least four of those oxidation states, and finally, (iv) must be able to promote the formation of an O–O bond.

The design of an ideal photoelectrosynthesis cell requires the incorporation of both oxidative and reductive elements, where efficient catalysis, light absorption and current matching will be essential for the overall device efficiency. Such devices are still far from being a reality, but the developments observed in recent years, particularly in regards to the molecular level understanding of the functional principles of PSII and the role of proton-coupled electron transfer in facilitating multiredox catalysis, and the somewhat successful synthesis of new models of the OEC, are encouraging and must be used as an impetus to push the research field of artificial photosynthesis/solar fuels even further.

References

1. G. F. Moore and G. W. Brudvig, *Annu. Rev. Condens. Matter Phys.*, 2011, **2**, 303–327.
2. I. McConnell, G. Li and G. W. Brudvig, *Chem. Biol.*, 2010, **17**, 434–447.
3. J. Barber, *Chem. Soc. Rev.*, 2009, **38**, 185–196.
4. V. Balzani, A. Credi and M. Venturi, *ChemSusChem*, 2008, **1**, 26–58.
5. J. H. Alstrum-Acevedo, M. K. Brennaman and T. J. Meyer, *Inorg. Chem.*, 2005, **44**, 6802–6827.
6. A. J. Bard and M. A. Fox, *Acc. Chem. Res.*, 1995, **28**, 141–145.

7. M. Yagi and M. Kaneko, *Chem. Rev.*, 2000, **101**, 21–36.
8. W. Rüttinger and G. C. Dismukes, *Chem. Rev.*, 1997, **97**, 1–24.
9. J. M. Mayer, I. J. Rhile, F. B. Larsen, E. A. Mader, T. F. Markle and A. G. DiPasquale, *Photosynth. Res.*, 2006, **87**, 3–20.
10. C. W. Cady, R. H. Crabtree and G. W. Brudvig, *Coord. Chem. Rev.*, 2008, **252**, 444–455.
11. H. Dau and I. Zaharieva, *Acc. Chem. Res.*, 2009, **42**, 1861–1870.
12. G. Renger and T. Renger, *Photosynth. Res.*, 2008, **98**, 53–80.
13. J. P. McEvoy and G. W. Brudvig, *Chem. Rev.*, 2006, **106**, 4455–4483.
14. J. S. Vrettos and G. W. Brudvig, *Philos. Trans. R. Soc., B*, 2002, **357**, 1395–1405.
15. G. T. Babcock, B. A. Barry, R. J. Debus, C. W. Hoganson, M. Atamian, L. McIntosh, I. Sithole and C. F. Yocum, *Biochemistry*, 1989, **28**, 9557–9565.
16. R. A. Binstead, B. A. Moyer, G. J. Samuels and T. J. Meyer, *J. Am. Chem. Soc.*, 1981, **103**, 2897–2899.
17. T. J. Meyer, *Acc. Chem. Res.*, 1989, **22**, 163–170.
18. L. Hammarstrom and S. Styring, *Philos. Trans. R. Soc., B*, 2008, **363**, 1283–1291.
19. J. Y. Fang and S. Hammes-Schiffer, *J. Chem. Phys.*, 1997, **107**, 8933–8939.
20. J. Y. Fang and S. Hammes-Schiffer, *J. Chem. Phys.*, 1997, **107**, 5727–5739.
21. J. Y. Fang and S. Hammes-Schiffer, *J. Chem. Phys.*, 1997, **106**, 8442–8454.
22. R. I. Cukier, *J. Phys. Chem.*, 1994, **98**, 2377–2381.
23. R. I. Cukier, *J. Phys. Chem.*, 1995, **99**, 16101–16115.
24. R. I. Cukier, *J. Phys. Chem.*, 1996, **100**, 15428–15443.
25. R. I. Cukier, *J. Phys. Chem. A*, 1999, **103**, 5989–5995.
26. R. I. Cukier and D. G. Nocera, *Annu. Rev. Phys. Chem.*, 1998, **49**, 337–369.
27. S. Hammes-Schiffer, *Acc. Chem. Res.*, 2001, **34**, 273–281.
28. S. Hammes-Schiffer, *ChemPhysChem*, 2002, **3**, 33–42.
29. R. I. Cukier, *Biochim. Biophys. Acta, Bioenerg.*, 2004, **1655**, 37–44.
30. S. Hammes-Schiffer and N. Iordanova, *Biochim. Biophys. Acta, Bioenerg.*, 2004, **1655**, 29–36.
31. J. M. Mayer, *Annu. Rev. Phys. Chem.*, 2004, **55**, 363–390.
32. J. M. Mayer and I. J. Rhile, *Biochim. Biophys. Acta, Bioenerg.*, 2004, **1655**, 51–58.
33. M. H. V. Huynh and T. J. Meyer, *Chem. Rev.*, 2007, **107**, 5004–5064.
34. J. Rosenthal and D. G. Nocera, *Acc. Chem. Res.*, 2007, **40**, 543–553.
35. C. Costentin, *Chem. Rev.*, 2008, **108**, 2145–2179.
36. S. Hammes-Schiffer, E. Hatcher, H. Ishikita, J. H. Skone and A. V. Soudackov, *Coord. Chem. Rev.*, 2008, **252**, 384–394.
37. S. J. Edwards, A. V. Soudackov and S. Hammes-Schiffer, *J. Phys. Chem. B*, 2009, **113**, 14545–14548.
38. S. Y. Reece and D. G. Nocera, *Annu. Rev. Biochem.*, 2009, **78**, 673–699.

39. P. E. M. Siegbahn and M. R. A. Blomberg, *Chem. Rev.*, 2010, **110**, 7040–7061.
40. J. J. Warren, T. A. Tronic and J. M. Mayer, *Chem. Rev.*, 2010, **110**, 6961–7001.
41. G. T. Babcock, M. Espe, C. Hoganson, N. LydakisSimantiris, J. McCracken, W. J. Shi, S. Styring, C. Tommos and K. Warncke, *Acta Chem. Scand.*, 1997, **51**, 533–540.
42. C. W. Hoganson and G. T. Babcock, *Science*, 1997, **277**, 1953–1956.
43. C. Tommos and G. T. Babcock, *Acc. Chem. Res.*, 1998, **31**, 18–25.
44. S. Hammes-Schiffer, *Proc. Natl. Acad. Sci. U. S. A.*, 2011, **108**, 8531–8532.
45. C. Venkataraman, A. V. Soudackov and S. Hammes-Schiffer, *J. Phys. Chem. C*, 2010, **114**, 487–496.
46. C. J. Fecenko, H. H. Thorp and T. J. Meyer, *J. Am. Chem. Soc.*, 2007, **129**, 15098–15099.
47. S. Hammes-Schiffer and A. A. Stuchebrukhov, *Chem. Rev.*, 2010, **110**, 6939–6960.
48. C. Venkataraman, A. V. Soudackov and S. Hammes-Schiffer, *J. Phys. Chem. C*, 2008, **112**, 12386–12397.
49. I. Navrotskaya, A. V. Soudackov and S. Hammes-Schiffer, *J. Chem. Phys.*, 2008, **128**, 244712.
50. C. Costentin, C. Louault, M. Robert and J.-M. Saveant, *Proc. Natl. Acad. Sci. U. S. A.*, 2009, **106**, 18143–18148.
51. C. Costentin, M. Robert and J. M. Saveant, *J. Am. Chem. Soc.*, 2006, **128**, 4552–4553.
52. C. Costentin, C. Louault, M. Robert and J.-M. Saveant, *J. Am. Chem. Soc.*, 2008, **130**, 15817–15819.
53. B. Auer, L. E. Fernandez and S. Hammes-Schiffer, *J. Am. Chem. Soc.*, 2011, **133**, 8282–8292.
54. J. M. Saveant, C. Costentin, M. Robert and C. Tard, *Angew. Chem., Int. Ed.*, 2010, **49**, 3803–3806.
55. H. Petek and J. Zhao, *Chem. Rev.*, 2010, **110**, 7082–7099.
56. L. A. Lyon and J. T. Hupp, *J. Phys. Chem. B*, 1999, **103**, 4623–4628.
57. S. R. Pendlebury, M. Barroso, A. J. Cowan, K. Sivula, J. Tang, M. Gratzel, D. Klug and J. R. Durrant, *Chem. Commun.*, 2011, **47**, 716–718.
58. J. Barber, *Biochim. Biophys. Acta, Bioenerg.*, 1998, **1365**, 269–277.
59. G. W. Brudvig, *Philos. Trans. R. Soc., B*, 2008, **363**, 1211–1219.
60. P. E. M. Siegbahn, *Acc. Chem. Res.*, 2009, **42**, 1871–1880.
61. C. Tommos, *Philos. Trans. R. Soc. London, B*, 2002, **357**, 1383–1394.
62. K. N. Ferreira, T. M. Iverson, K. Maghlaoui, J. Barber and S. Iwata, *Science*, 2004, **303**, 1831–1838.
63. A. Guskov, J. Kern, A. Gabdulkhakov, M. Broser, A. Zouni and W. Saenger, *Nat. Struct. Mol. Biol.*, 2009, **16**, 334–342.
64. Y. Umena, K. Kawakami, J.-R. Shen and N. Kamiya, *Nature*, 2011, **473**, 55–60.
65. F. Rappaport and B. A. Diner, *Coord. Chem. Rev.*, 2008, **252**, 259–272.

66. F. Rappaport, M. Guergova-Kuras, P. J. Nixon, B. A. Diner and J. Lavergne, *Biochemistry*, 2002, **41**, 8518–8527.
67. B. Kok, B. Forbush and M. Mcgloin, *Photochem. Photobiol.*, 1970, **11**, 457–475.
68. H. Dau, C. Limberg, T. Reier, M. Risch, S. Roggan and P. Strasser, *ChemCatChem*, 2010, **2**, 724–761.
69. B. A. Diner and F. Rappaport, *Annu. Rev. Plant Biol.*, 2002, **53**, 551–580.
70. M. R. A. Blomberg, P. E. M. Siegbahn, S. Styring, G. T. Babcock, B. Akermark and P. Korall, *J. Am. Chem. Soc.*, 1997, **119**, 8285–8292.
71. M. L. Gilchrist, J. A. Ball, D. W. Randall and R. D. Britt, *Proc. Natl. Acad. Sci. U. S. A.*, 1995, **92**, 9545–9549.
72. L. I. Krishtalik, *Biochim. Biophys. Acta*, 1986, **849**, 162–171.
73. L. I. Krishtalik, *Bioelectrochem. Bioenerg.*, 1990, **23**, 249–263.
74. S. Gerken, K. Brettel, E. Schlodder and H. T. Witt, *FEBS Lett.*, 1988, **237**, 69–75.
75. M. Sjodin, T. Irebo, J. E. Utas, J. Lind, G. Merenyi, B. Akermark and L. Hammarstrom, *J. Am. Chem. Soc.*, 2006, **128**, 13076–13083.
76. L. Hammarstrom and S. Styring, *Energy Environ. Sci.*, 2011, **4**, 2379–2388.
77. H. Dau and M. Haumann, *Coord. Chem. Rev.*, 2008, **252**, 273–295.
78. F. Rappaport and J. Lavergne, *Biochim. Biophys. Acta, Bioenerg.*, 2001, **1503**, 246–259.
79. W. Lubitz, E. J. Reijerse and J. Messinger, *Energy Environ. Sci.*, 2008, **1**, 15–31.
80. M. Hambourger, G. F. Moore, D. M. Kramer, D. Gust, A. L. Moore and T. A. Moore, *Chem. Soc. Rev.*, 2009, **38**, 25–35.
81. A. L. Moore, M. Kloz, S. Pillai, G. Kodis, D. Gust, T. A. Moore, R. van Grondelle and J. T. M. Kennis, *J. Am. Chem. Soc.*, 2011, **133**, 7007–7015.
82. A. L. Moore, Y. Terazono, G. Kodis, K. Bhushan, J. Zaks, C. Madden, T. A. Moore, G. R. Fleming and D. Gust, *J. Am. Chem. Soc.*, 2011, **133**, 2916–2922.
83. D. Gust, T. A. Moore and A. L. Moore, *Acc. Chem. Res.*, 1993, **26**, 198–205.
84. L. C. Sun, L. Hammarstrom, B. Akermark and S. Styring, *Chem. Soc. Rev.*, 2001, **30**, 36–49.
85. F. E. Osterloh, *Chem. Mater.*, 2008, **20**, 35–54.
86. A. Kay, I. Cesar and M. Gratzel, *J. Am. Chem. Soc.*, 2006, **128**, 15714–15721.
87. A. Harriman, I. J. Pickering, J. M. Thomas and P. A. Christensen, *J. Chem. Soc. Faraday Trans. 1*, 1988, **84**, 2795–2806.
88. A. Harriman, J. M. Thomas and G. R. Millward, *New J. Chem.*, 1987, **11**, 757–762.
89. F. Jiao and H. Frei, *Energy Environ. Sci.*, 2010, **3**, 1018–1027.
90. A. Kudo and Y. Miseki, *Chem. Soc. Rev.*, 2009, **38**, 253–278.

91. K. Maeda and K. Domen, *J. Phys. Chem. Lett.*, 2010, **1**, 2655–2661.
92. R. Breslow, *Acc. Chem. Res.*, 1995, **28**, 146–153.
93. M. Yagi, Y. Takahashi, I. Ogino and M. Kaneko, *J. Chem. Soc., Faraday Trans.*, 1997, **93**, 3125–3127.
94. J. L. Dempsey, A. J. Esswein, D. R. Manke, J. Rosenthal, J. D. Soper and D. G. Nocera, *Inorg. Chem.*, 2005, **44**, 6879–6892.
95. C. J. Gagliardi, J. W. Jurss, H. H. Thorp and T. J. Meyer, *Inorg. Chem.*, 2011, **50**, 2076–2078.
96. A. J. Morris, G. J. Meyer and E. Fujita, *Acc. Chem. Res.*, 2009, **42**, 1983–1994.
97. D. Gust, T. A. Moore, A. L. Moore, A. A. Krasnovsky, P. A. Liddell, D. Nicodem, J. M. Degraziano, P. Kerrigan, L. R. Makings and P. J. Pessiki, *J. Am. Chem. Soc.*, 1993, **115**, 5684–5691.
98. D. Gust, T. A. Moore, A. L. Moore, L. Leggett, S. Lin, J. M. Degraziano, R. M. Hermant, D. Nicodem, P. Craig, G. R. Seely and R. A. Nieman, *J. Phys. Chem.*, 1993, **97**, 7926–7931.
99. A. L. Moore, M. Gervaldo, P. A. Liddell, G. Kodis, B. J. Brennan, C. R. Johnson, J. W. Bridgewater, T. A. Moore and D. Gust, *Photochem. Photobiol. Sci.*, 2010, **9**, 890–900.
100. A. L. Moore, A. E. Keirstead, J. W. Bridgewater, Y. Terazono, G. Kodis, S. Straight, P. A. Liddell, T. A. Moore and D. Gust, *J. Am. Chem. Soc.*, 2010, **132**, 6588–6595.
101. A. Magnuson, H. Berglund, P. Korall, L. Hammarstrom, B. Akermark, S. Styring and L. C. Sun, *J. Am. Chem. Soc.*, 1997, **119**, 10720–10725.
102. A. Magnuson, Y. Frapart, M. Abrahamsson, O. Horner, B. Akermark, L. C. Sun, J. J. Girerd, L. Hammarstrom and S. Styring, *J. Am. Chem. Soc.*, 1999, **121**, 89–96.
103. F. Jiao and H. Frei, *Angew. Chem., Int. Ed.*, 2009, **48**, 1841–1844.
104. M. W. Kanan and D. G. Nocera, *Science*, 2008, **321**, 1072–1075.
105. M. W. Kanan, Y. Surendranath and D. G. Nocera, *Chem. Soc. Rev.*, 2009, **38**, 109–114.
106. M. W. Kanan, J. Yano, Y. Surendranath, M. Dinca, V. K. Yachandra and D. G. Nocera, *J. Am. Chem. Soc.*, 2010, **132**, 13692–13701.
107. J. G. McAlpin, Y. Surendranath, M. Dinca, T. A. Stich, S. A. Stoian, W. H. Casey, D. G. Nocera and R. D. Britt, *J. Am. Chem. Soc.*, 2010, **132**, 6882.
108. Y. Surendranath, M. Dinca and D. G. Nocera, *J. Am. Chem. Soc.*, 2009, **131**, 2615.
109. M. D. Symes, Y. Surendranath, D. A. Lutterman and D. G. Nocera, *J. Am. Chem. Soc.*, 2011, **133**, 5174.
110. R. Brimblecombe, A. Koo, G. C. Dismukes, G. F. Swiegers and L. Spiccia, *J. Am. Chem. Soc.*, 2010, **132**, 2892–2894.
111. G. C. Dismukes, R. Brimblecombe, G. A. N. Felton, R. S. Pryadun, J. E. Sheats, L. Spiccia and G. F. Swiegers, *Acc. Chem. Res.*, 2009, **42**, 1935–1943.

112. S. W. Gersten, G. J. Samuels and T. J. Meyer, *J. Am. Chem. Soc.*, 1982, **104**, 4029–4030.

113. J. A. Gilbert, D. S. Eggleston, W. R. Murphy, D. A. Geselowitz, S. W. Gersten, D. J. Hodgson and T. J. Meyer, *J. Am. Chem. Soc.*, 1985, **107**, 3855–3864.

114. H. Yamada, W. F. Siems, T. Koike and J. K. Hurst, *J. Am. Chem. Soc.*, 2004, **126**, 9786–9795.

115. R. A. Binstead, C. W. Chronister, J. F. Ni, C. M. Hartshorn and T. J. Meyer, *J. Am. Chem. Soc.*, 2000, **122**, 8464–8473.

116. Z. Deng, H.-W. Tseng, R. Zong, D. Wang and R. Thummel, *Inorg. Chem.*, 2008, **47**, 1835–1848.

117. F. Liu, J. J. Concepcion, J. W. Jurss, T. Cardolaccia, J. L. Templeton and T. J. Meyer, *Inorg. Chem.*, 2008, **47**, 1727–1752.

118. K. Nagoshi, S. Yamashita, M. Yagi and M. Kaneko, *J. Mol. Catal. A: Chem.*, 1999, **144**, 71–76.

119. J. L. Cape, S. V. Lymar, T. Lightbody and J. K. Hurst, *Inorg. Chem.*, 2009, **48**, 4400–4410.

120. J. J. Concepcion, J. W. Jurss, M. K. Brennaman, P. G. Hoertz, A. O. T. Patrocinio, N. Y. M. Iha, J. L. Templeton and T. J. Meyer, *Acc. Chem. Res.*, 2009, **42**, 1954–1965.

121. J. J. Concepcion, J. W. Jurss, J. L. Templeton and T. J. Meyer, *Proc. Natl. Acad. Sci. U. S. A.*, 2008, **105**, 17632–17635.

122. J. W. Jurss, J. C. Concepcion, M. R. Norris, J. L. Templeton and T. J. Meyer, *Inorg. Chem.*, 2010, **49**, 3980–3982.

123. R. A. Binstead, B. A. Moyer, G. J. Samuels and T. J. Meyer, *J. Am. Chem. Soc.*, 1981, **103**, 2897.

124. G. C. Dismukes, R. Brimblecombe, G. A. N. Felton, R. S. Pryadun, J. E. Sheats, L. Spiccia and G. F. Swiegers, *Acc. Chem. Res.*, 2009, **42**, 1935.

125. S. Bernhard, N. D. McDaniel, F. J. Coughlin and L. L. Tinker, *J. Am. Chem. Soc.*, 2008, **130**, 210–217.

126. J. A. Seabold and K.-S. Choi, *Chem. Mater.*, 2011, **23**, 1105–1112.

127. K. J. McDonald and K.-S. Choi, *Chem. Mater.*, 2011, **23**, 1686–1693.

128. E. R. Young, D. G. Nocera and V. Bulovic, *Energy Environ. Sci.*, 2010, **3**, 1726–1728.

129. D. K. Zhong, J. W. Sun, H. Inumaru and D. R. Gamelin, *J. Am. Chem. Soc.*, 2009, **131**, 6086–6087.

130. M. Barroso, A. J. Cowan, S. R. Pendlebury, M. Gratzel, D. R. Klug and J. R. Durrant, *J. Am. Chem. Soc.*, 2011, **133**, 14868.

131. D. K. Zhong and D. R. Gamelin, *J. Am. Chem. Soc.*, 2010, **132**, 4202–4207.

132. H. J. M. de Groot, *Appl. Magn. Reson.*, 2010, **37**, 497–503.

133. W. J. Youngblood, S.-H. A. Lee, Y. Kobayashi, E. A. Hernandez-Pagan, P. G. Hoertz, T. A. Moore, A. L. Moore, D. Gust and T. E. Mallouk, *J. Am. Chem. Soc.*, 2009, **131**, 926–927.

134. C. J. Gagliardi, B. C. Westlake, C. A. Kent, J. J. Paul, J. M. Papanikolas and T. J. Meyer, *Coord. Chem. Rev.*, 2010, **254**, 2459–2471.

135. B. C. Westlake, M. K. Brennaman, J. J. Concepcion, J. J. Paul, S. E. Bettis, S. D. Hampton, S. A. Miller, N. V. Lebedeva, M. D. E. Forbes, A. M. Moran, T. J. Meyer and J. M. Papanikolas, *Proc. Natl. Acad. Sci.*, 2011, **108**, 8554–8558.

136. E. E. Benson, C. P. Kubiak, A. J. Sathrum and J. M. Smieja, *Chem. Soc. Rev.*, 2009, **38**, 89–99.

137. K. Heinze, K. Hempel and M. Beckmann, *Eur. J. Inorg. Chem.*, 2006, 2040–2050.

138. A. F. Heyduk, A. M. Macintosh and D. G. Nocera, *J. Am. Chem. Soc.*, 1999, **121**, 5023–5032.

139. J. Rosenthal, J. Bachman, J. L. Dempsey, A. J. Esswein, T. G. Gray, J. M. Hodgkiss, D. R. Manke, T. D. Luckett, B. J. Pistorio, A. S. Veige and D. G. Nocera, *Coord. Chem. Rev.*, 2005, **249**, 1316–1326.

140. Z. Y. Bian, K. Sumi, M. Furue, S. Sato, K. Koike and O. Ishitani, *Inorg. Chem.*, 2008, **47**, 10801–10803.

141. K. Tanaka and D. Ooyama, *Coord. Chem. Rev.*, 2002, **226**, 211–218.

142. K. Toyohara, H. Nagao, T. Mizukawa and K. Tanaka, *Inorg. Chem.*, 1995, **34**, 5399–5400.

143. E. E. Barton, D. M. Rampulla and A. B. Bocarsly, *J. Am. Chem. Soc.*, 2008, **130**, 6342–6344.

144. K. Frese and S. Leach, *J. Electrochem. Soc.*, 1985, **132**, 259–272.

145. J. Li and G. Prentice, *J. Electrochem. Soc.*, 1997, **144**, 4284–4288.

146. K. Ohkawa, Y. Noguchi, S. Nakayama, K. Hashimoto and A. Fujishima, *J. Electroanal. Chem.*, 1994, **367**, 165–173.

147. J. Popic, M. AvramovIvic and N. Vukovic, *J. Electroanal. Chem.*, 1997, **421**, 105–110.

148. J. Qu, X. Zhang, Y. Wang and C. Xie, *Electrochim. Acta*, 2005, **50**, 3576–3580.

149. D. Summers, S. Leach and K. Frese, *J. Electroanal. Chem.*, 1986, **205**, 219–232.

150. M. Watanabe, M. Shibata, A. Kato, M. Azuma and T. Sakata, *J. Electrochem. Soc.*, 1991, **138**, 3382–3389.

151. B. Aurian-Blajeni, M. Halmann and J. Manassen, *Sol. Energy Mater.*, 1983, **8**, 425–440.

152. D. Canfield and J. K. W. Frese, *J. Electrochem. Soc.*, 1983, **130**, 1772–1774.

153. M. Halmann, *Nature*, 1978, **275**, 115–116.

154. K. Ogura and M. Takagi, *J. Electroanal. Chem.*, 1986, **206**, 209–216.

Subject Index